AF368956

Remote Sensing in Coastline Detection

Remote Sensing in Coastline Detection

Editor

Donatella Dominici

MDPI • Basel • Beijing • Wuhan • Barcelona • Belgrade • Manchester • Tokyo • Cluj • Tianjin

Editor
Donatella Dominici
DICEAA, Dept. of Civil,
Environmental Engineering and
Architecture
Italy

Editorial Office
MDPI
St. Alban-Anlage 66
4052 Basel, Switzerland

This is a reprint of articles from the Special Issue published online in the open access journal *Journal of Marine Science and Engineering* (ISSN 2077-1312) (available at: https://www.mdpi.com/journal/jmse/special_issues/RS_coastline).

For citation purposes, cite each article independently as indicated on the article page online and as indicated below:

LastName, A.A.; LastName, B.B.; LastName, C.C. Article Title. *Journal Name* **Year**, *Article Number*, Page Range.

ISBN 978-3-03936-836-5 (Hbk)
ISBN 978-3-03936-837-2 (PDF)

Cover image courtesy of Donatella Dominici.

Contents

About the Editor . . . . **vii**

Donatella Dominici and Sara Zollini
Remote Sensing in Coastline Detection
Reprinted from: *J. Mar. Sci. Eng.* **2020**, *8*, 498, doi:10.3390/jmse8070498 . . . . **1**

**Marco Anzidei, Fawzi Doumaz, Antonio Vecchio, Enrico Serpelloni, Luca Pizzimenti,
Riccardo Civico, Michele Greco, Giovanni Martino and Flavio Enei**
Sea Level Rise Scenario for 2100 A.D. in the Heritage Site of Pyrgi (Santa Severa, Italy)
Reprinted from: *J. Mar. Sci. Eng.* **2020**, *8*, 64, doi:10.3390/jmse8020064 . . . . **3**

Antonio Zanutta, Alessandro Lambertini and Luca Vittuari
UAV Photogrammetry and Ground Surveys as a Mapping Tool for Quickly Monitoring
Shoreline and Beach Changes
Reprinted from: *J. Mar. Sci. Eng.* **2020**, *8*, 52, doi:10.3390/jmse8010052 . . . . **21**

**Sara Zollini, Maria Alicandro, María Cuevas-González, Valerio Baiocchi, Donatella
Dominici and Paolo Massimo Buscema**
Shoreline Extraction Based on an Active Connection Matrix (ACM) Image Enhancement
Strategy
Reprinted from: *J. Mar. Sci. Eng.* **2020**, *8*, 9, doi:10.3390/jmse8010009 . . . . **37**

Maria Alicandro, Valerio Baiocchi, Raffaella Brigante and Fabio Radicioni
Automatic Shoreline Detection from Eight-Band VHR Satellite Imagery
Reprinted from: *J. Mar. Sci. Eng.* **2019**, *7*, 459, doi:10.3390/jmse7120459 . . . . **53**

Arthur Trembanis, Alimjan Abla, Ken Haulsee and Carter DuVal
Benthic Habitat Morphodynamics-Using Remote Sensing to Quantify Storm-Induced Changes
in Nearshore Bathymetry and Surface Sediment Texture at Assateague National Seashore
Reprinted from: *J. Mar. Sci. Eng.* **2019**, *7*, 371, doi:10.3390/jmse7100371 . . . . **67**

**Francesco Immordino, Mattia Barsanti, Elena Candigliota, Silvia Cocito, Ivana Delbono and
Andrea Peirano**
Application of Sentinel-2 Multispectral Data for Habitat Mapping of Pacific Islands: Palau
Republic (Micronesia, Pacific Ocean)
Reprinted from: *J. Mar. Sci. Eng.* **2019**, *7*, 316, doi:10.3390/jmse7090316 . . . . **97**

**Giovanni Pugliano, Umberto Robustelli, Diana Di Luccio, Luigi Mucerino, Guido Benassai
and Raffaele Montella**
Statistical Deviations in Shoreline Detection Obtained with Direct and Remote Observations
Reprinted from: *J. Mar. Sci. Eng.* **2019**, *7*, 137, doi:10.3390/jmse7050137 . . . . **113**

About the Editor

Donatella Dominici is a Professor of Geomatic at the University of L'Aquila, and obtained her a PhD in Geodetic and Topographic Sciences from the Engineering College of Bologna. Between 1994 and 1998, she held a permanent researcher position in the disciplinary group ICAR/06—Topography and Cartography—at the Institute of Topography, Geodesy and Mineral Geophysics, of the Engineering Faculty of Bologna. From 1998 to the present day, she had been an Associate Professor in the disciplinary group ICAR/06—Topography and Cartography—at the University of L'Aquila. Her areas of research include traditional and GNSS surveying, GNSS data processing, UAV photogrammetry, remote sensing, as well as the use of the Artificial Intelligence Algorithm in environmental studies. She has worked on and managed a significant number of funded national and international projects, as well as start-up projects (GITAIS srl). These projects covered research area such as GNSS data processing, NRTK implementation, geomatic techniques applied in early warming study, SAR and optical data processing, UAV photogrammetry in post-earthquake scenario, coastline detection with remote sensing data, and the use of the Artificial Intelligence Algorithm to improve data analysis in an environmental and cultural heritage scenario. She has established the Geomatic Laboratory of the University of L'Aquila, which is now a facility enabling smart cooperation with other universities and national organizations. She has been and continues to be an active member of a number of departmental groups at the University of L'Aquila along with various external University interview panels, committees and academic boards. She has organized, participated in and chaired a number of national and international conferences throughout her career. She has extensive undergraduate and graduate level of teaching experience and has organised and taught at a number of summer schools. She has also taught a number of professional refresher courses. She has successfully supervised over seventy undergraduate theses along with six PhD projects. She has more than one hundred publications.

Journal of
*Marine Science
and Engineering*

Editorial

Remote Sensing in Coastline Detection

Donatella Dominici * and Sara Zollini

DICEAA-Department of Civil, Construction-Architecture and Environmental Engineering,
University of L'Aquila, Via Gronchi 18, 67100 L'Aquila, Italy; sara.zollini@graduate.univaq.it
* Correspondence: donatella.dominici@univaq.it

Received: 1 June 2020; Accepted: 24 June 2020; Published: 7 July 2020

"Is beach erosion a natural cycle or is it getting worse with rising sea levels?" [1]

Coastal zones are some of the most populated and developed areas in the world.

They have rich biodiversity and it is said that more than 45% of the world's population lives there [2].

Coastal areas provide many resources but they are very vulnerable environments. Coastal hazards (e.g., typhoons/cyclones/hurricanes, storm surges, tsunamis) represent significant threats to the population, infrastructure and to the coasts themselves. There are also other hazards which are not visible or produce long-term effects, such as rising sea levels and coastal erosion.

Beach erosion is defined as the removal of sand from a beach to deeper water offshore or alongshore into inlets, tidal shoals and bays [1]. This is caused by various factors, including natural and anthropological factors such as the simple inundation of the land by rising sea levels resulting from the melting of the polar ice caps.

Climate change is one of the main reasons for this, but it is not the only one. Beaches are greatly influenced by the frequency and magnitude of storms along a particular shoreline.

Data on shoreline changes for the period between 1984 and 2016 (33 years) show that 24% of the world's sandy beaches are eroding at rates exceeding 0.5 m/yr, while 28% are accreting and 48% are stable. Moreover, the majority of the sandy shorelines in marine protected areas are eroding [3].

It is clear that the coastal environment needs to be protected, not only for heritage but also to preserve human life.

For the management of coastal areas, prevention assumes a central role. The protection of coastal green belts as well as the shoreline through coastal dykes or other hard infrastructures should be based on a carefully performed risk analysis. On the other hand, "softer" components, such as coastal watching and monitoring, are also important to aid the preservation of these areas [2]. In this sense, remote sensing takes on a fundamental role in "watching" and detecting coastal changes over time.

With the latest generation of high and very high resolution satellite images, it is possible to monitor a wide area of land at relatively low costs. Of course, there are many methods that can be used and that has been used in the literature to detect the coastal environment, like GNSS (Global Navigation Satellite System), UAV (Unmanned Aerial Vehicle) photogrammetry, video systems, traditional surveys using total station, levelling and so on, but remote sensing, compared to the others, is quicker, the images cover a larger portion of the territory, there is no need to go physically to the place under investigation and there is no need for an "ad hoc" flight. It is obvious that the synergy of all these techniques would allow researchers to reach more complete results; indeed, in most applications, satellite remote sensed images are not used alone. However, thanks to the arrangement of high-performance image analysis techniques, it is nowadays possible to at least obtain initial results about the problem to be solved.

To conclude, the main goal is to find an automated and replicable technique to evaluate the spatial and temporal evolution of alterations due to natural and anthropogenic events, especially for large areas, so that prompt action can be initiated.

I am grateful to all the colleagues who played a role in the drafting of this Special Issue [4–10], as they have allowed a good overview of all multiscale remote sensing techniques (high resolution images, photogrammetry, SAR (Synthetic Aperture Radar), GNSS, etc.) and a whole array of methods and techniques for the processing, analysis and discussion of multitemporal remotely sensed data.

Author Contributions: D.D. and S.Z. contributed equally to this work. Both authors have read and agreed to the published version of the manuscript.

Funding: This research received no external funding.

Acknowledgments: I would like to thank MDPI and above all Esme Wang for her patience and the support.

Conflicts of Interest: The authors declare no conflict of interest.

References

1. 'What Causes Beach Erosion?', Scientific American. Available online: https://www.scientificamerican.com/article/what-causes-beach-erosion/ (accessed on 21 April 2020).
2. Rajib, S. 'Chapter 1-Global Coasts in the Face of Disasters'. In *Coastal Management*; Krishnamurthy, R.R., Jonathan, M.P., Srinivasalu, S., Glaeser, B., Eds.; Elsevier: Amsterdam, The Netherlands, 2019; pp. 1–4.
3. Luijendijk, A.; Hagenaars, G.; Ranasinghe, R.; Baart, F.; Donchyts, G.; Aarninkhof, S. 'The State of the World's Beaches'. *Sci. Rep.* **2018**, *8*, 6641. [CrossRef] [PubMed]
4. Anzidei, M.; Doumaz, F.; Vecchio, A.; Serpelloni, E.; Pizzimenti, L.; Civico, R.; Greco, M.; Martino, G.; Enei, F. Sea Level Rise Scenario for 2100 A.D. in the Heritage Site of Pyrgi (Santa Severa, Italy). *J. Mar. Sci. Eng.* **2020**, *8*, 64. [CrossRef]
5. Zanutta, A.; Lambertini, A.; Vittuari, L. UAV Photogrammetry and Ground Surveys as a Mapping Tool for Quickly Monitoring Shoreline and Beach Changes. *J. Mar. Sci. Eng.* **2020**, *8*, 52. [CrossRef]
6. Zollini, S.; Alicandro, M.; Cuevas-González, M.; Baiocchi, V.; Dominici, D.; Buscema, P.M. Shoreline Extraction Based on an Active Connection Matrix (ACM) Image Enhancement Strategy. *J. Mar. Sci. Eng.* **2020**, *8*, 9. [CrossRef]
7. Alicandro, M.; Baiocchi, V.; Brigante, R.; Radicioni, F. Automatic Shoreline Detection from Eight-Band VHR Satellite Imagery. *J. Mar. Sci. Eng.* **2019**, *7*, 459. [CrossRef]
8. Trembanis, A.; Abla, A.; Haulsee, K.; DuVal, C. Benthic Habitat Morphodynamics-Using Remote Sensing to Quantify Storm-Induced Changes in Nearshore Bathymetry and Surface Sediment Texture at Assateague National Seashore. *J. Mar. Sci. Eng.* **2019**, *7*, 371. [CrossRef]
9. Immordino, F.; Barsanti, M.; Candigliota, E.; Cocito, S.; Delbono, I.; Peirano, A. Application of Sentinel-2 Multispectral Data for Habitat Mapping of Pacific Islands: Palau Republic (Micronesia, Pacific Ocean). *J. Mar. Sci. Eng.* **2019**, *7*, 316. [CrossRef]
10. Pugliano, G.; Robustelli, U.; Di Luccio, D.; Mucerino, L.; Benassai, G.; Montella, R. Statistical Deviations in Shoreline Detection Obtained with Direct and Remote Observations. *J. Mar. Sci. Eng.* **2019**, *7*, 137. [CrossRef]

Journal of
*Marine Science
and Engineering*

Article

Sea Level Rise Scenario for 2100 A.D. in the Heritage Site of Pyrgi (Santa Severa, Italy)

Marco Anzidei [1,*], Fawzi Doumaz [1], Antonio Vecchio [2,3], Enrico Serpelloni [1], Luca Pizzimenti [1], Riccardo Civico [1], Michele Greco [4], Giovanni Martino [4] and Flavio Enei [5]

[1]   Istituto Nazionale di Geofisica e Vulcanologia, 56126 Pisa, Italy; fawzi.doumaz@ingv.it (F.D.);
      enrico.serpelloni@ingv.it (E.S.); luca.pizzimenti@ingv.it (L.P.); riccardo.civico@ingv.it (R.C.)
[2]   Radboud Radio Lab, Department of Astrophysics/IMAPP, Radboud University— Nijmegen,
      6500GL Nijmegen, The Netherlands; antonio.vecchio@obspm.fr
[3]   Lesia Observatoire de Paris, Université PSL, CNRS, Sorbonne Université, Université de Paris,
      92195 Meudon, France
[4]   School of Engineering, Università della Basilicata, 85100 Potenza, Italy; michele.greco@unibas.it (M.G.);
      ing.gmartino@gmail.com (G.M.)
[5]   Museo del Mare e della Navigazione Antica, 00058 Santa Severa, Italy; muspyrgi@tiscali.it
*   Correspondence: marco.anzidei@ingv.it

Received: 29 November 2019; Accepted: 15 January 2020; Published: 21 January 2020

Abstract: Sea level rise is one of the main risk factors for the preservation of cultural heritage sites located along the coasts of the Mediterranean basin. Coastal retreat, erosion, and storm surges are posing serious threats to archaeological and historical structures built along the coastal zones of this region. In order to assess the coastal changes by the end of 2100 under the expected sea level rise of about 1 m, we need a detailed determination of the current coastline position based on high resolution Digital Surface Models (DSM). This paper focuses on the use of very high-resolution Unmanned Aerial Vehicles (UAV) imagery for the generation of ultra-high-resolution mapping of the coastal archaeological area of Pyrgi, Italy, which is located near Rome. The processing of the UAV imagery resulted in the generation of a DSM and an orthophoto with an accuracy of 1.94 cm/pixel. The integration of topographic data with two sea level rise projections in the Intergovernmental Panel on Climate Change (IPCC) AR5 2.6 and 8.5 climatic scenarios for this area of the Mediterranean are used to map sea level rise scenarios for 2050 and 2100. The effects of the Vertical Land Motion (VLM) as estimated from two nearby continuous Global Navigation Satellite System (GNSS) stations located as close as possible to the coastline are included in the analysis. Relative sea level rise projections provide values at 0.30 ± 0.15 cm by 2050 and 0.56 ± 0.22 cm by 2100 for the IPCC AR5 8.5 scenarios and at 0.13 ± 0.05 cm by 2050 and 0.17 ± 0.22 cm by 2100, for the IPCC Fifth Assessment Report (AR5) 2.6 scenario. These values of rise correspond to a potential beach loss between 12.6% and 23.5% in 2100 for Representative Concentration Pathway (RCP) 2.6 and 8.5 scenarios, respectively, while, during the highest tides, the beach will be provisionally reduced by up to 46.4%. In higher sea level positions and storm surge conditions, the expected maximum wave run up for return time of 1 and 100 years is at 3.37 m and 5.76 m, respectively, which is capable to exceed the local dune system. With these sea level rise scenarios, Pyrgi with its nearby Etruscan temples and the medieval castle of Santa Severa will be exposed to high risk of marine flooding, especially during storm surges. Our scenarios show that suitable adaptation and protection strategies are required.

Keywords: sea level rise; coastlines; 2100; storm surges; heritage sites; Pyrgi; Mediterranean; UAV; DSM

---

1. Introduction

Observational and instrumental data collected worldwide since the last two-three centuries show that the global sea level is continuously rising with an accelerated trend in recent years, which coincides with the rise in global temperatures [1]. Global mean sea level is expected to rise by about 75 to 200 cm by 2100 in the worst scenarios [2–5], i.e., the most serious effects of climate change that might occur in future decades. These values will be even larger in subsiding coasts of the Mediterranean, entailing widespread environmental changes, coastal retreat, marine flooding, and loss of land, which will be disadvantages for human activities. The sea level rise will amplify the impacts exerted by a multitude of hazards (i.e., storm surges, flooding, coastal erosion, and tsunamis) on the infrastructure and building integrity, people safety, economic assets, and cultural heritage.

Therefore, it is important to mitigate these risks by providing multi-temporal scenarios of expected inland extension of marine flooding as a consequence of the sea level rise for a cognizant coastal management [6–8].

This is particularly true for the Mediterranean region where ancient civilizations were born and developed along its coasts [9,10]. A large number of heritage sites are located at the waterfront or very close to the sea level and are exposed to marine flooding under the effects of ongoing climate change. A large part of these sites, which are dated back to the Greek, Roman, and Medieval ages, are exposed at increasing risks to coastal hazards that are related to a sea-level rise [10].

Aerial photogrammetric surveys performed by small Unmanned Aerial Vehicles (UAVs) can provide accurate topography at low costs and, in short time, for small areas with respect to conventional aerial surveys [11]. Results, when analyzed in combination with sea level rise projections and vertical land movements (VLM), can support the realization of the expected sea-level rise and storm surge scenarios for future decades [12].

In this study, we show an effective application for the coastal archaeological area of Pyrgi, Italy, which is near Rome (Figure 1). High resolution maps of expected flooded areas and coastal positions for 2100, even in storm surge conditions, are reported in this paper.

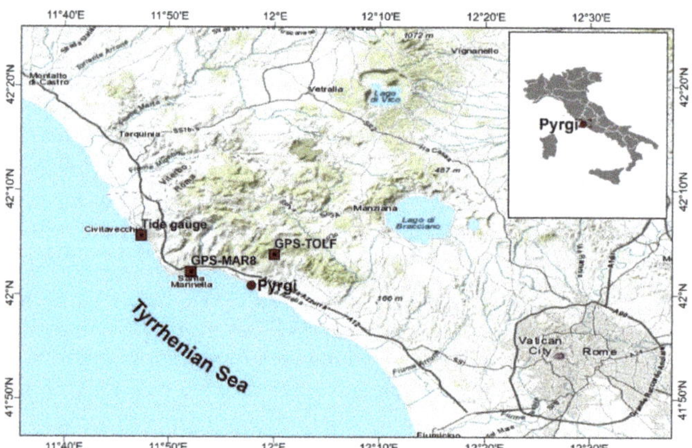

Figure 1. The investigated area of Pyrgi with the location of the GNSS stations of MAR8 and TOLF and the nearby tide gauge station of Civitavecchia.

Our approach is to apply a multidisciplinary methodology previously tuned in the savemedcoasts project (www.savemedcoasts.eu), which includes topography, geodesy, sea level data, and climatic projections to estimate realistic sea level rise scenarios for targeted coastal areas. Our approach provides a useful analytical tool to identify the best adaptation and defense strategies against the sea level rise impact, and to protect heritage sites. The results help decision-makers in the selection of the best practical actions aimed at preserving the archaeological and historical sites located in coastal areas that are subjected to sea level-related risks. The proposed methodology can be exported in other areas of the Mediterranean region and beyond its borders.

2. Geo-Archaeological Setting

The heritage site of Pyrgi is located along the coasts of Northern Latium, between the villages of Santa Severa and Cerveteri, which is about 50 km north of Rome (Italy) (Figure 1). The area includes the Castle of Santa Severa that, with Pyrgi, is one of the most important heritage sites of the Tyrrhenian coast. The area has been settled since the V-IV millennium B.C. [13] and continuously developed during the Neolithic Age, during the Bronze Age (II millennium B.C.), and during the Iron Age (IX-VIII century B.C.), thanks to its good environmental conditions. In the Etruscan phase (VII-IV century B.C.), Pyrgi was the port of the ancient Etruscan city of Caere and played an important role in the maritime commerce being frequented by Greeks and Phoenicians ships. The area includes a sanctuary and the temples of Eileithyia-Leukothea and Apollo, Cavatha, Suri, and the Etruscan Uni analogue to the Phoenician Astarte [13].

After the Romanization of this area (III century B.C.), Pyrgi became a maritime colony and a tall fortress surrounded by a polygonal wall was built on part of the Etruscan settlement.

During the Roman imperial age, the city of Pyrgi continued to be frequented until the 5th-6th century A.D. when a Byzantine castrum was built on its remains with the early Christian church of Santa Severa inside.

Later, the medieval and renaissance village became a large farm located in a strategic position between the main harbors of Rome and Civitavecchia [13–16]. In terms of its geological setting, the coast of Pyrgi belongs to the roman co-magmatic province [17] that underwent major volcanic activity during the Plio-Pleistocene era (Figure 2). The surface geology is characterized by a bedrock belonging to the allochthonous Outer Ligurids [18], represented by the Tolfa Formation, spanning from the Late Cretaceous to Palaeogene [19], defined as the Pietraforte formation and "comprehensive succession" [20]. The latter consists of a series of marl limestone and grey marl beds that outcrop between Rome and Civitavecchia underlying the Pietraforte unit. Biogenic sandstones of the Early-Middle Pleistocene Panchina Formation partly overlies the Pietraforte [21].

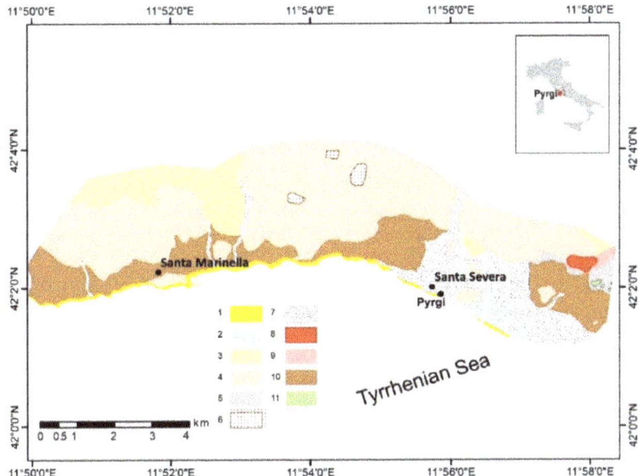

Figure 2. Simplified geological map of the study area (from http://dati.lazio.it/catalog/it/dataset/carta-geologica-informatizzata-regione-lazio-25000). Legend: (1) anthropic debris, (2) coastal and marsh sands and recent dunes, (3) calcareous marls and clays, (4) flysch, (5) clays with chalks, (6) landslide deposits, (7) alluvial deposits, (8) lavas, (9) travertines, (10) sandy deposits, and (11) sandy deposits of marine facies.

The Holocene deposits are represented by travertines, slope debris, alluvial, weathering deposits, gravels, and sandy beaches [22]. The neotectonic of this sector of the Tyrrhenian coast of Italy is marked by the elevation of the MIS 5.5 marine terraces that show stability and a weak uplift of the inland sector for the last 124 years [23,24], which is related to magmatic injections under the Vulsini and Sabatini volcanic complexes [24]. Reference [25] underlines that the long-term uplift may not be an appropriate description for all the past two millennia since some weak subsidence may have occurred at Pyrgi and along the nearby coasts.

3. Methods

We applied a multidisciplinary approach using coastal topography, geodesy, and climatic-driven estimates of the sea-level rise to provide maps of flooding scenarios for the year 2100 A.D. for the coast of Pyrgi. Our study consists of three main steps: (1) the realization of UAV surveys to obtain an ultra-high-resolution orthophoto and a DSM model of the coastal area to map the current and the projected coastline positions including sea level data in the analysis, (2) the estimation of the current vertical land movements from the analysis of geodetic data from the nearest GPS stations, and, (3) by combining these data with the regional IPCC-AR5 projections (RCP-2.6 and RCP-8.5 scenarios), the calculation of the upper bounds of the expected sea levels for the targeted epochs of 2050 and 2100 A.D. and the corresponding expected inland extent of the marine flooding and shoreline positions were calculated. Lastly, storm surge scenarios were implemented for return times of 1 and 100 years, for sea level rise conditions.

4. Digital Terrain Model Reconstruction

To realize the ultra-high-resolution Digital Surface Model (DSM), an aerial photogrammetric survey was performed using a radio-controlled multi-rotor Da Jiang Innovation (DJI) Phantom 4pro UAV system, equipped with a high resolution lightweight digital camera, to capture a set of aerial images of the investigated area (Table 1).

Table 1. Survey features.

Survey and Camera Features			
Number of images	306	Camera stations	286
Flying altitude	79.8 m	Tie points	1,158,646
Ground resolution	1.94 cm/pix	Projections	3,891,753
Coverage area	0.226 km^2	Reprojection error	0.528 pix
Camera model	FC6310 (8.8 mm)	Focal length	8.8 mm
Camera resolution	5472 × 3648	Pixel size	2.41 × 2.41 μm

The UAV was controlled by an autopilot system using waypoints previously planned by PIX4D® capture IOs App as a Ground Control Station system. To optimize the photogrammetric spatial resolution and coverage of the surveyed area, a constant altitude of 70 m was maintained during the flight and 306 partly overlapping (70% of longitudinal and 70% lateral overlapping between subsequent photos) aerial digital photos were acquired during three successful flights of 13, 11 and 15 min of duration each. To scale the aerial images, we used a dual-frequency geodetic Global Positioning System Real Time Kinematic (GPS/RTK) receiver STONEX S900A® to measure the coordinates of a set of reference Ground Control Points (GCPs) falling in the investigated areas. GCPs positions were estimated in real time by the RTK technique with about 1 to 2 cm of accuracy, with respect to the reference GPS station TOLF. The latter is part of the GNSS network of the Istituto Nazionale di Geofisica e Vulcanologia (INGV) [26] and also pertains to the Leica SmartNet ItalPoS network (https://hxgnsmartnet.com/en-gb/) for real-time positioning services. We used the Agisoft PhotoScan®

software package (http://www.agisoft.com) based on the Structure-from-Motion photogrammetry technique [27] to process the acquired georeferenced images. The analysis included: (1) camera alignment with image position and orientation, (2) generation of a dense points cloud, and (3) generation of an orthophoto covering a land surface of 0.226 km^2 with a ground resolution of 1.94 cm/pixel as well as the DSM creation in the WGS84-UTM32 coordinate system.

The obtained ortho-rectified images (orthophotos) and digital elevation models were also managed by Global Mapper® and ESRI ArcGis® software to create cross sections, slope maps, surfaces, and coastline positions as well as calculate the dimension of the potential flooded areas. The extracted Digital Surface Model has a resolution of 15.5 cm/pixel and a point density of 41.5 points/m^2

As targets for GCPs measured by GPS/RTK, we used a set of i) natural markers belonging to fixed structures (e.g., the center of manholes, wall and sidewalk corners, small structures, or large stones), and ii) mobile targets for the time of the flight, such as thin metal crosses of 60×60 cm of size, deployed in the investigated area. All these GCPs were chosen as areas recognizable on the images during data analysis.

In total, 22 Ground Control Points (GCPs) and Check Points (CPs) were used to geo-reference the orthophoto with the ground control point toolset of AGISOFT photo scan (Figure 3). The 3D coordinates of these points have been estimated with a mean RMS of 0.6 and 0.9 cm for the horizontal and vertical components, respectively, and were used to evaluate the vertical accuracy of our final DSM. Figure 3 shows the orthophoto while Figure 4 shows the DSM.

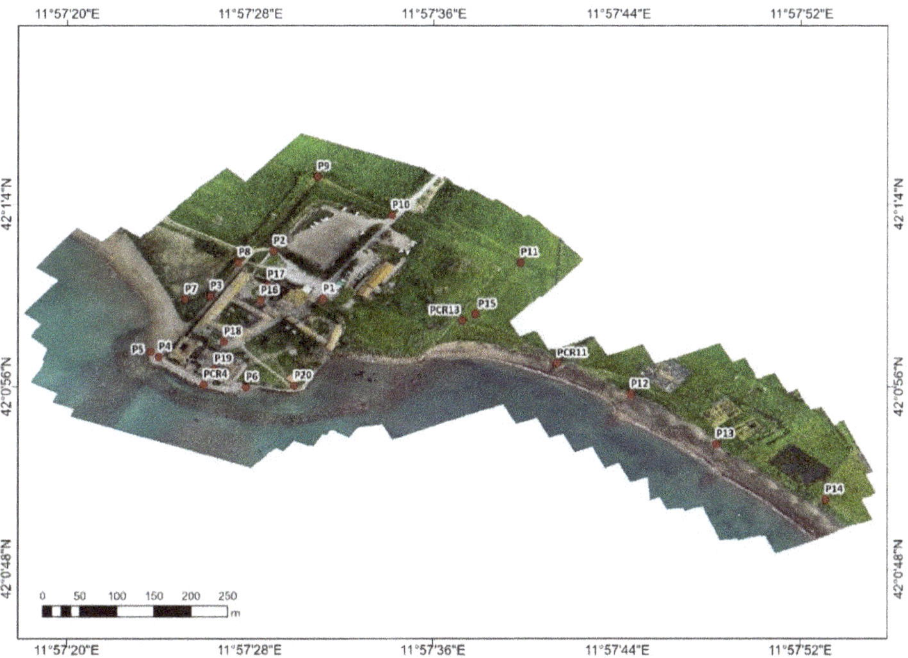

Figure 3. The orthophoto with the Ground Control Points (red dots) position used during the UAV surveys.

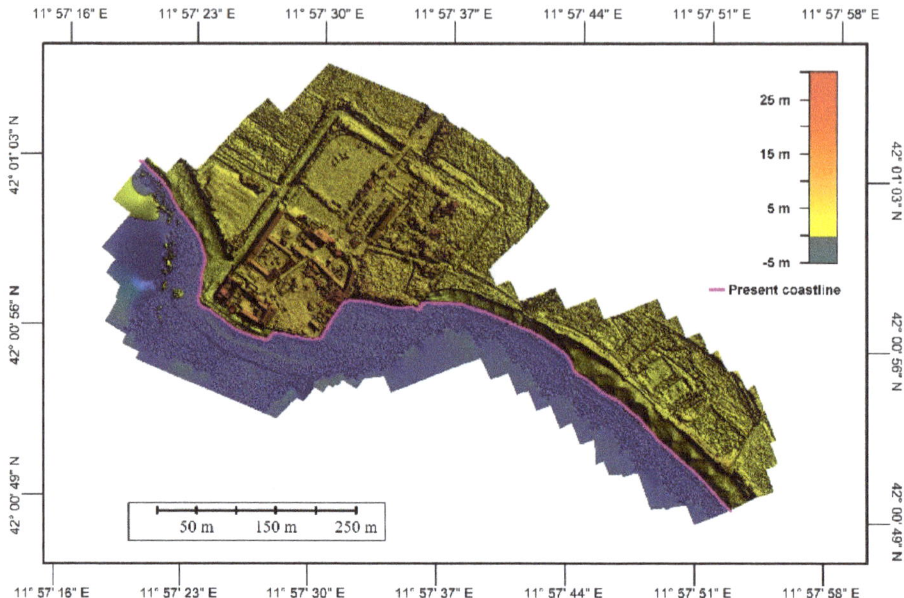

Figure 4. The Digital Surface Model (DSM) of the coastal sector of Pyrgi, from the analysis of the aerial photos.

5. Tidal Correction and Coastline Position

The coast of Pyrgi, similar to most of the coasts of the Mediterranean Sea, is characterized by a micro-tidal environment and tides are generally in the range of ±30 cm. Only in the Gulf of Gabes (Tunisia) and the North Adriatic Sea (Italy), tides may reach amplitudes up to about 2 m [27–30]. We used the tidal data collected by the Italian tidal network managed by the Italian Institute for Environmental Protection and Research (ISPRA) (data are freely available at www.mareografico.it) at the sea level station of Civitavecchia (located at LAT 42°05′38.25″, LON 11°47′22.73′), which is placed near Pyrgi (Figure 1), to estimate the tide level (TL) and the local mean sea level (LMSL) at the time of the UAV surveys (26 April, 2019 at 07:30 or 05:30 UTC). We considered the complete time series to account for a long-term linear trend, representative of the mean sea level during UAV surveys, with respect to which the TL is defined. This tide gauge station shows a valid recording period of about eight years (2011–2019) with a sea level trend of 0.25 ± 0.1 mm/a, which is calculated from a linear fit on the monthly data (Figure 5a). From the data analysis, a mean tide amplitude of ~35 cm and a maximum tidal range up to ~60 cm has been estimated (Figure 5b). To define a reference level for elevation data, the mean sea level was computed propagating the linear trend from the time of surveys, assuming the mean sea level as a reference value for the year 2018. A mean sea level of 4.8 ± 11.8 cm above the topographic benchmark was estimated from the hourly tidal data. Since the UAV surveys have been performed during a high tide of 4 cm, as shown by the tidal recordings (www.mareografico.it), then the reference sea level at the time of surveys corresponds to a position of only 0.8 cm above the mean sea level for 2018.

This value is negligible compared to the DSM accuracy, so we did not consider it for the analysis of sea level rise scenarios. Lastly, we used the local mean sea level calculated at the tide station of Civitavecchia as elevation data, given the small distance between this location and Pyrgi. The obtained value was used to define the position of the coastline during the surveys and the one expected from the sea level rise scenarios to 2100 A.D.

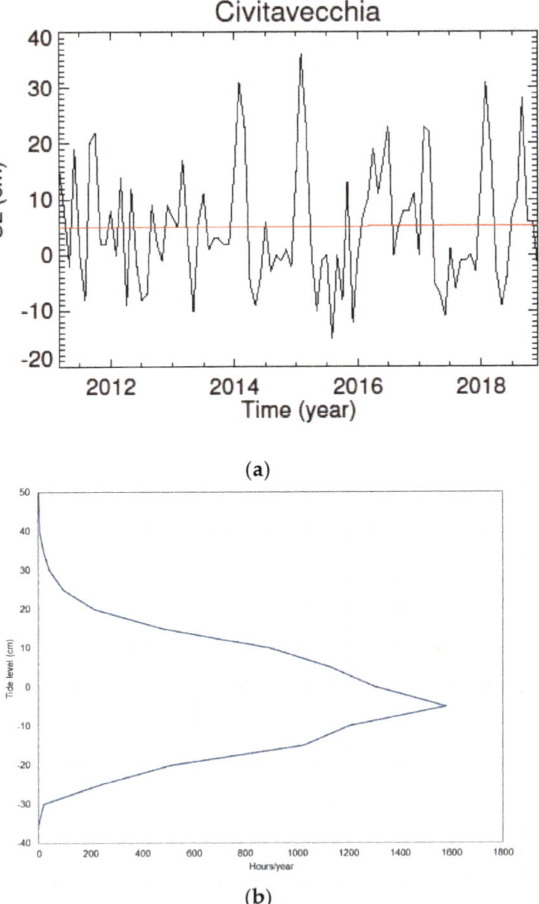

Figure 5. Sea level data analysis for the tide gauge of Civitavecchia, which is located a few km north of Pyrgi (see Figure 1 for location). (**a**) Monthly data of sea level recordings collected in the time span 2011–2019 (about nine years). The red line is the linear fit of the sea level trend at 0.25 ± 0.1 mm/yr. (**b**) Statistical diagram of sea level heights (cm) versus time (hours/year). The values of sea level height frequency are reported during one year, including maximum sea level heights of about 45 cm that may exceed the tide amplitudes. These can be related to storm surge events that occur only a few hours in a year, when water is pushed from the sea onto the land due to a temporary decrease in atmospheric pressure and wind.

We preferred to adopt this local vertical datum instead of the value of the geoid elevation for the Italian region provided by the International Service for the Geoid-ISG-ITG2009 (http://www.isgeoid.polimi.it/Geoid/Europe/Italy/ITG2009_g.html) [31] since it is an independent and more accurate elevation datum. The International Service for the Geoid (ISG) estimates a geoid height in the Italian Geoid (ITG) 2009 for the coastline of Pyrgi at 48.319 m. This elevation corresponds to the contour line equal to zero in the Italian height reference frame. The DSM height reference frame is 0.42 m, as estimated by GPS/RTK data at a GCP located at the sea level along the coastline. Its elevation was corrected for the tidal range at the time of the surveys. The reference mean sea level estimated by the tidal analysis provided a local mean sea level with an uncertainty of ±11.8 cm.

6. Vertical Land Motion (VLM) at Pyrgi

The current rate of VLM at Pyrgi was estimated by the analysis of the available GPS data collected at the nearest GNSS stations of TOLF, belonging the INGV Rete Integrata Nazionale GPS (RING) network (DOI:10.13127/RING) and MAR8, belonging to the Topcon-NetGeo network (http://www.netgeo.it). These stations, which are located at about 6.5 km and 0.3 km of distance from the study site, respectively (Figure 1), have a robust time series that span for 2004–2019 for TOLF (15.23 years) and 2012–2019 for MAR8 (7.38 years) (Figure 6).

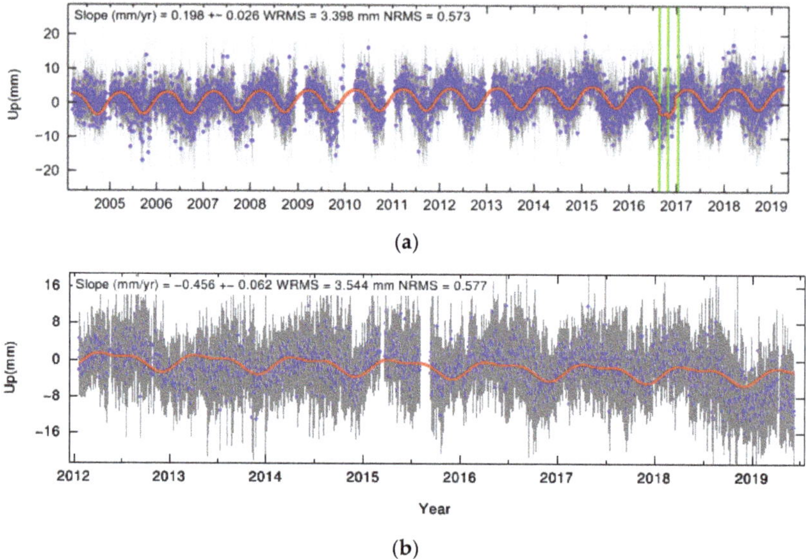

Figure 6. The vertical components (UP) of the time series of the GNSS station for (**a**) TOLF (time span 2004–2019, about 15 years) and (**b**) MAR8 (time span 2012–2019), both located near Pyrgi (see Figure 1 for location).

GPS data analysis has been carried out following the procedures already described in Reference [32] and updated in Reference [33] by adopting a three-step procedure using the GAMIT/GLOBK V10.7 [34] and QOCA software. This is part of a continental-scale GPS solution including >3000 stations [34]. The daily positions of TOLF and MAR8 have been estimated in the GPS realization of the ITRF2008 frame [35], i.e., the IGb08 reference. The position time series have been analyzed in order to estimate and correct offsets due to station equipment changes while, simultaneously, estimating annual and semi-annual periodic signals and a linear velocity term, whereas velocity uncertainties have been estimated adopting a power law + white noise stochastic model, as in Reference [36]. The results show that both sites are relatively stable in the IGb08 reference frame with a vertical velocity of −0.061 ± 0.135 mm/year for TOLF and −0.456 ± 0.344 mm/year for MAR8. We remark that uncertainties associated on the vertical velocities are about ±0.5 mm/year and are barely significant in view of unresolved questions about the GPS reference frame stability and additional factors [30].

In addition to GPS data, the tectonic stability of this region is also inferred from the low level of seismicity deduced from historical data [37] and instrumental recordings of earthquakes (www.ingv.it), which do not report the occurrence of significant events for the last 3000 years BP.

Lastly, assuming that the area will continue in the near future to have the same tectonic trend shown in the past, it is reasonable to neglect the contribution of VLM in the sea level rise projections and flooding scenarios for 2100 A.D.

7. Relative Sea-Level Rise Projections and Flooding Scenarios for 2050 and 2100 A.D.

To estimate the sea-level rise for 2050 and 2100 A.D. at Pyrgi, we referred to the regional IPCC AR5 sea-level projections discussed in the Fifth Assessment Report of the IPCC-AR5 [3], www.ipcc.ch (data available from the Integrated Climate data Center-ICDC of the University of Hamburg, http://icdc.cen.uni hamburg.de/1/daten/ocean/ar5-slr.html). These data consist of the sea-level ensemble mean values and upper/lower 90% confidence bounds of the sea level on a global grid (spatial resolution 1° × 1°), obtained by adding the contributions of several geophysical sources driving long-term sea-level changes: (1) the thermosteric/dynamic contribution (from 21 CMIP5 coupled atmosphere-ocean general circulation models AOGCMs), (2) the surface mass balance and dynamic ice sheet contributions from Greenland and Antarctica, (3) the glacier and land water storage contributions, (4) the Glacial Isostatic Adjustment (GIA), and (5) the inverse barometer effect [1]. Projections, which are based on two different Representative Concentration Pathways RCP 2.6 and RCP 8.5 while providing the least and most amounts of future sea level rise, respectively, were used. The IPCC regional sea-level rate at the grid point closest to the location of the tide gauge station (Civitavecchia) was considered. By accounting for VLM from GPS data, very high-resolution DSM and regional IPCC sea level projections at the grid point corresponding to the investigated area, the first marine flooding scenarios for Pyrgi for 2050 and 2100 A.D. have been realized. To include the VLM effect in sea-level projections, we substituted the modelled GIA contribution to the IPCC rates with the GPS vertical velocities, which includes both GIA and tectonic components. Uncertainties for the sea-level estimations were calculated by combining lower and upper sea level bounds from IPCC projection and errors from GPS measurements. In any case, given the tectonic stability of the area, the VLM have a null contribution in the analysis. The relative sea-level rise in RCP2.6 and RCP8.5 scenarios at 2050 and 2100 A.D. with respect to the chosen reference epoch 2017 are shown in Figure 7, and numerical values are reported in Table 2.

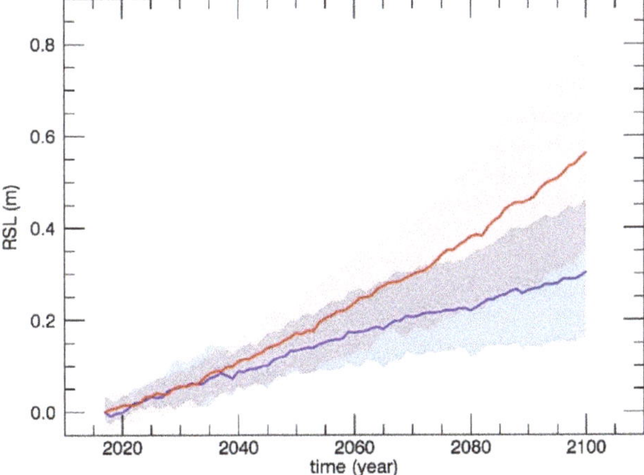

Figure 7. Relative sea level with respect to the 2017 level as obtained from the regional IPCC sea-level projections, AR5 RCP2.6 (blue line), and RCP8.5 (red line) for a null VLM. Color bands represent the 90% confidence interval. The small-scale variations observed in the data are related to the ocean component contribution accounting for the effects of dynamic Sea Surface Height (SSH), the global thermosteric SSH anomaly, and inverse barometer effects (Church et al., 2013a, b, http://icdc.cen.unihamburg.de/). Given the vertical tectonic stability of the area, the VLM have a null contribution in the projections.

Table 2. Relative Sea Level Rise (RSLR, cm) above the current mean sea level at Pyrgi for 2050 and 2100 for the IPCC 2.6 and 8.5 climatic scenarios, in the mean and maximum high tide conditions. Given the vertical tectonic stability of the area, the VLM were not considered in the analysis.

Sea Level 2050 IPCC 8.5 (cm)			Sea Level 2050 IPCC 2.6 (cm)		
RSLR	RSLR + Mean high tide (30 cm)	RSLR + Max high tide (45 cm)	RSLR	RSLR + Mean high tide (30 cm)	RSLR + Max high tide (45 cm)
17 ± 0.07	47	62	13	43	58
Sea level 2100 IPCC 8.5 (cm)			Sea level 2100 IPCC 2.6 (cm)		
RSLR	RSLR + Mean high tide (30 cm)	RSLR + Max high tide (45 cm)	RSLR	RSLR + Mean high tide (30 cm)	RSLR + Max high tide (45 cm)
56 ± 0.22	86	101	30	60	75

The projected coastline positions for 2100 A.D., corresponding to the RCP2.6 and RCP8.5 scenarios, are obtained from the DSM for the sea levels listed in Table 2 and shown in Figure 8. The computed and the represented scenarios correspond to the local mean sea level (estimated with the uncertainty of ±11.8 cm), and are obtained by neglecting the periodical contribution due to diurnal and semidiurnal tides. To account for it, in order to estimate the maximum inundation scenarios, the time series of daily hydrometric sea levels at Civitavecchia tidal station from January 2010 to December 2018 were included in the analysis. The average half amplitude of the daily tides was estimated as high as about 30 cm (Figure 5b). Consequently, in the RCP 8.5 scenario for 2100, if we take into account both the sea-level rise (56 cm) and the mean daily tides (30 cm), we can infer a maximum water level of 86 cm. Since the highest sea levels may reach 45 cm for a few hours a year during temporary meteorological conditions when a decrease in atmospheric pressure and wind push water from the sea onto the land (Figure 5b), the maximum expected water level may reach up to 101 cm above the current level.

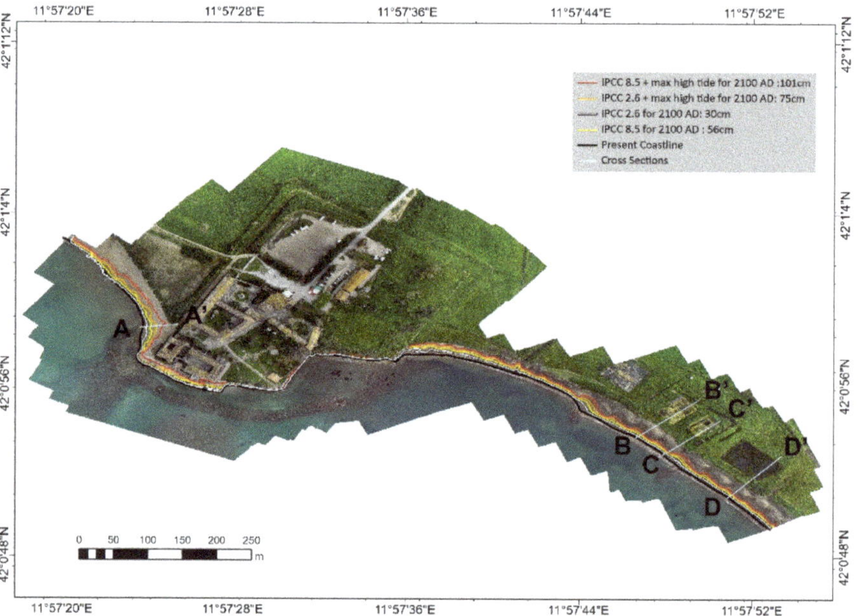

Figure 8. Projected coastline positions for 2100 A.D. for the 2.6 and 8.5 IPCC scenarios (see legend for details). Coastline positions include the uncertainty of ±11.8 cm in the reference mean sea level position as estimated from the analysis of tide gauge data. The cross sections along the shore are also shown in the figure.

For the RCP 2.6 scenario for 2100, considering a sea-level rise of 30 cm and the same mean daily tides (30 cm), a maximum sea-level rise of 60 cm can be inferred. During the rare maximum highest sea levels (45 cm) (Figure 5b), the maximum expected water level would reach 75 cm above the 2019 level.

The sea level rise scenarios for 2100 A.D. depicts the impact of the marine flooding for the coast surrounding the castle of Santa Severa, the temples of Pyrgi, and nearby beaches. The cross sections shown in Figure 9 traced along a direction that is normal to the coast and across the archaeological area of Pyrgi, which highlights the presence of a soft dune system running parallel to the shore. This system is well developed south of the castle for the whole bay of Pyrgi, while, in the north side, it is buried by the modern buildings of the Santa Severa village.

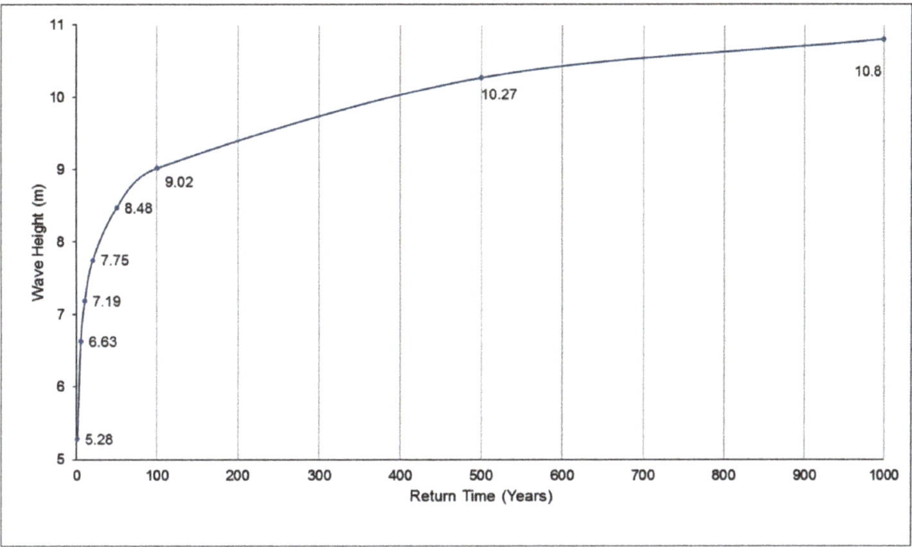

Figure 9. Return time (years) values of the significant wave heights (m).

Since the Etruscan time (2450 years BP), the coastline has retreated at about 80 m, at a mean rate of 3.3 cm/yr [38]. The temple of Pyrgi was originally placed at more than 120–125 m from the present-day shore and well protected from the sea by the dune system. Today, it is placed at about 40–45 m from the coastline and is undergoing water intrusion and frequent flooding, especially during storm surges (Figures 8 and 9). The proximity of the sea and the accelerated erosion of the dune system, which is highly exposed to waves especially during storm surges, are causing a continuous coastal erosion and beach retreat. Based on our sea level rise projections, a beach retreat of about 10 m is expected for 2100 A.D. (see cross sections B, C, and D in Figures 8 and 9).

This rapid retreat will accelerate the erosion process and the likely dismantling of the soft dune system with the consequent direct exposure of the temples to the sea. At the same time, the beach located along the north-west side of the castle, which is lacking a dune system and characterized by a gentle slope, is expected to retreat up to about 25 m (see cross section A), exposing the foot of the castle to severe erosion and flooding.

Lastly, Table 3 shows the expected flooded areas in 2050 and 2100 A.D. for the RCP 8.5 and RCP 2.6 climatic scenarios, together with the corresponding percentage of beach loss, which may reach up to 46% during the highest tides.

Table 3. IPCC scenarios, relative sea level rise projections for 2100, expected land loss (m^2), and percentage of land loss with respect to the current surface for the given Relative Sea Level Rise (RSLR) projections. Given the vertical tectonic stability of the area, the VLM were considered in the analysis.

IPCC Scenario	RSLR 2100 (cm)	Land Loss (m^2)	Percent of Land Loss (16113.55 m^2)
IPCC 2.6	17 ± 0.07	2033.33	12.6
IPCC 2.6 + max high tide	75 ± 0.22	5387.87	33.4
IPCC 8.5	56 ± 0.22	3790.47	23.5
IPCC 8.5 + max high tide	101 ± 0.22	7480.02	46.4

8. Storm Surge Scenarios

To evaluate the storm surge scenario for Pyrgi, a local climate wave assessment was performed using the Weibull distribution and the Equivalent Triangular Storm model originally proposed by Boccotti [39]. The return values of the significant wave heights and related peak wave periods (Table 4) have been evaluated on the parameters of the Weibull distribution [40] using data collected at the wave buoy of Ponza (wave data are freely available at http://dati.isprambiente.it/), which is located in the central Tyrrhenian sea in front of the coasts of Latium (at coordinates Latitude 40.867°N and Longitude 12.950°E).

Table 4. Return values R_T (years) of the significant wave height H_S (m) and the related peak wave period T_P (s).

R_T (Years)	H_S (m)	T_P (s)
1	5.28	9.80
5	6.63	10.98
10	7.19	11.43
20	7.75	11.87
50	8.48	12.41
100	9.02	12.80
500	10.27	13.66
1000	10.80	14.01

The climate wave data for ordinary and extreme storm wave conditions in terms of the return period (R_T) for R_T = 1 year and R_T = 100 years in a sea level rise condition for 2100 in the RCP8.5 climatic scenario reported in Tables 2 and 3 were used to estimate wave setup and wave run-up during the event for both return times (Figure 10).

In the present analysis, wind set up was neglected and the precautionary assumption that the direction of wave travel does not produce refraction was considered. The expression derived by Weggel [41] and the relation of Komar and Gaughan estimated breaker depth index and breaker height index [42]. The breaker type was correlated with the surf similarity parameter, which allows the run-up estimation by the use of the predictive equations [43,44]. Based on our topographic surveys, a mean beach slope of 3.7% for the investigated coast, was assumed for the wave run-up assessment.

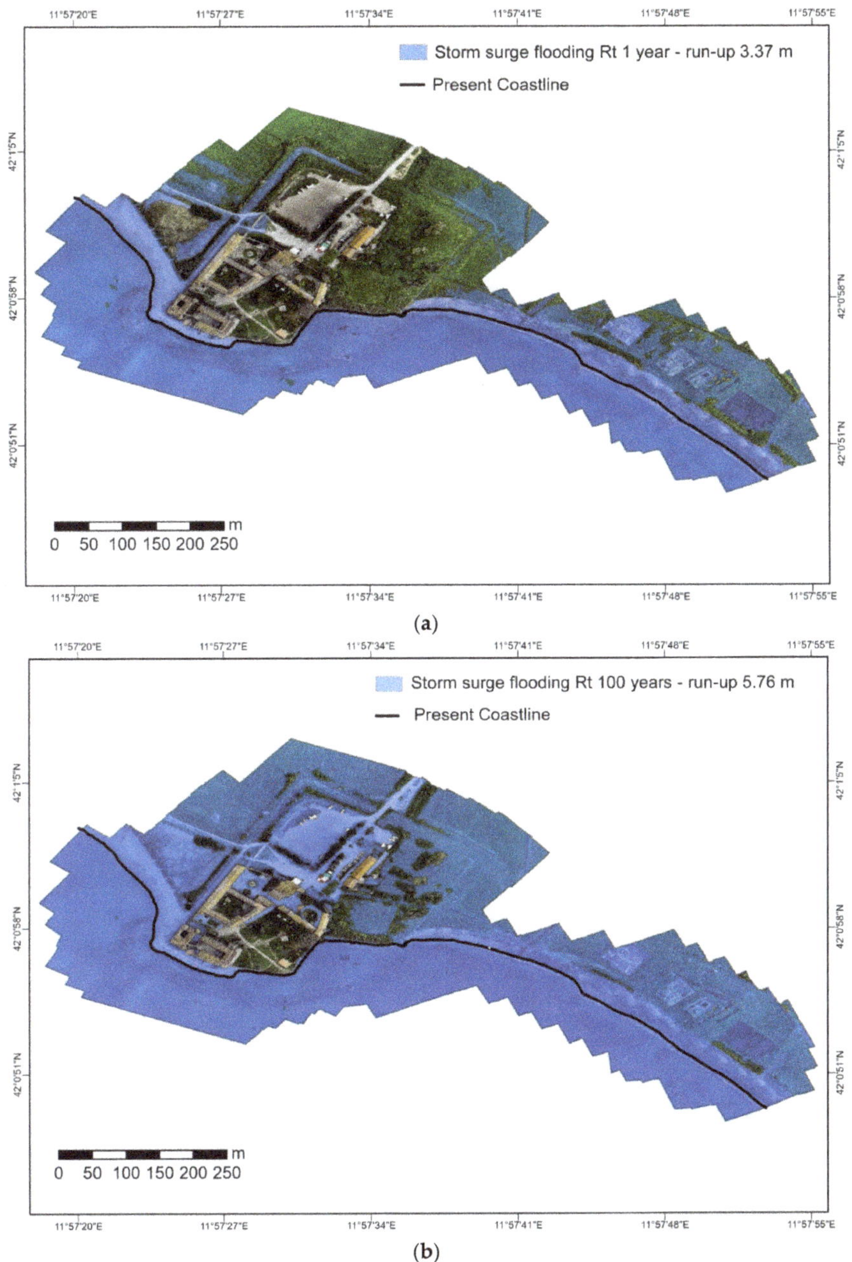

Figure 10. Projected flooded areas corresponding to the maximum run-up levels for storm surges with return times for: (**a**) 1 year (in blue at 3.37 m) and (**b**) 100 years (in blue at 5.76 m). The coastline position on 26 April 2019 is in black, which is estimated with an uncertainty of ±11.8 cm from the analysis of tide gauge data.

The results of the wave run-up analysis in ordinary and extreme storm conditions (Table 5 and Figure 11), were estimated at 3.37 m and 5.76 m for return time of 1 and 100 years, respectively.

Table 5. Values of maximum R_{max} and medium R_{med} wave run-up (m) estimated in the offshore wave conditions (H_S and T_P), beach slope (s m/m), breaking depth d_b (m), and self-similarity parameter (ξ_0).

R_T (Years)	H_s (m)	T_p (sec)	s(m/m)	d_b (m)	ξ_0	R_{max} (m)	R_{med} (m)
1	5.28	9.80	0.037	6.27	0.187	3.37	1.46
100	9.02	12.80	0.037	10.71	0.187	5.76	2.50

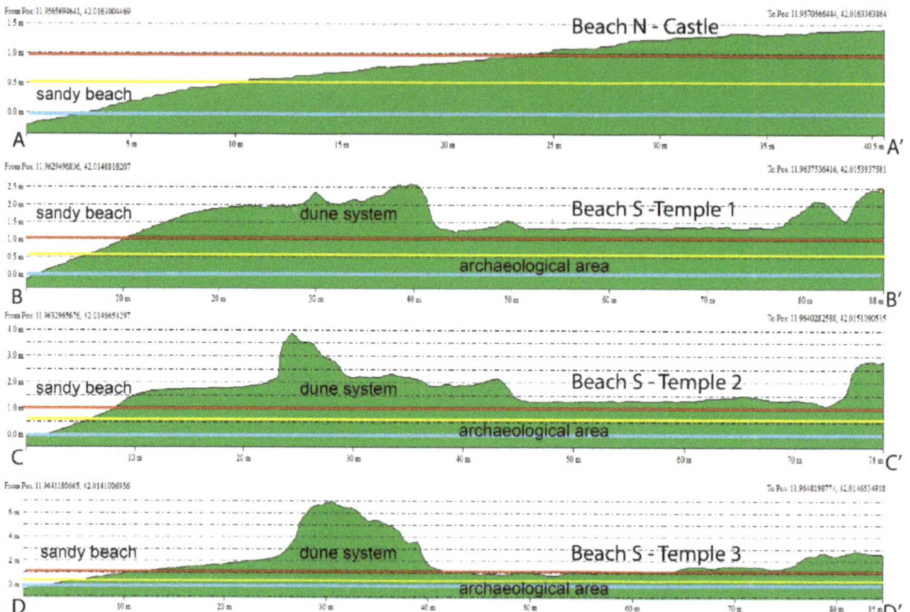

Figure 11. Cross sections of the coastline. See Figure 7 for cross section positions. A-A' (Beach North–Castle) north of the castle of Santa Severa; B-B', C-C' and D-D' (Beach South–Temples) across the beach and the three structures of the temple of Pyrgi. The expected sea levels for 2100 in the RCP8.5 scenario and in the maximum RCP8.5 high tide condition, are shown by the yellow and red lines, respectively. Light blue is the current mean sea level position. For storm surge scenarios for return time 1 and 100 years in sea level rise conditions, the area is almost entirely flooded with the possible exception across D-D', which remains protected by the high dune system. Projected coastlines include the uncertainty of ±11.8 cm in the mean sea level position as estimated from the analysis of tide gauge data.

9. Discussion and Conclusions

In this study, we assessed the marine flooding scenarios for 2050 and 2100 for the relevant heritage sites of Pyrgi on the basis of (a) a high-resolution DSM, (b) rates of VLM from GPS data, and (c) tidal data from the nearby tide gauge and the RCP 2.6 and 8.5 climatic scenarios released by the IPCC. The results highlight how the flooding might impact both the beach and archaeological sites with a potential local sea-level rise for the IPCC 8.5 scenario of about 56 cm for 2100 A.D. During the highest sea levels estimated by the tidal data at the nearby Civitavecchia station, the sea-level could potentially reach more than 86 cm and up to 101 cm. In addition, 2450 years ago, the shoreline was extended seaward of about 80 m more than today while, during the ancient Roman period, the sea level was at about 1.2 m below the current position, as estimated by the fish tank data [34]. Hence, a continuous sea level rise is occurring since the last millennia with an acceleration that started about 100 ± 53 years ago, at the beginning of the Industrial Era [25]. Due to the potential significant impacts on both the coast and the heritage sites of the Temple of Pyrgi and the Santa Severa Castle with a beach retreat up to about 25 m, the expected scenario reported in this study can support adaptation plans at different time scales, which are in agreement with the Protocol on Integrated Coastal Zone Management (ICZM, https://eur-lex.europa.eu/eli/prot/2009/89/oj) in the Mediterranean.

Our results detail previous studies for the Italian [7–9,12,45–47] and the Mediterranean [10,48] regions and can contribute to raise awareness of policymakers and heritage managers toward the coastal hazard by highlighting the need to adapt actions to protect Pyrgi from marine flooding and erosion under the current conditions due to the expected sea level rise and storm surge scenarios.

We remark that the unceasing coastal erosion since the Etruscan time [38] and the retreat of the soft cliffs characterizing the Pyrgi coastline mostly occurs during high energy marine events. During storm surges, the waves approaching the coast from the Northwest, West, and Southwest marine sectors are particularly dangerous for Pyrgi. The waves approaching the coast with fetches up to about 400 km are weakly slowed down by the seafloor morphology, which results in an increase of wave energy in sea level rise conditions [49] and leads to a maximum wave run-up of 5.76 m for a return time of 100 years. The impact of extreme events may be significantly amplified by the effects of climate change. In this scenario, the intensities of the strongest future storms will exceed the strength of any in the past in the Mediterranean Sea and the oceans [50–53]. The effects of the sea-level rise in the projected scenarios include severe coastal erosion and potential enhanced damages to the heritage site of Pyrgi, which, as many other coastal areas of the Mediterranean, deserves rapid actions for their future preservation.

Author Contributions: Conceptualization, M.A. and A.V. Methodology, M.A. Software; validation, M.A., A.V., and M.G. Formal analysis, A.V., F.D., E.S., M.G., and G.M. Investigation, L.P., R.C., M.A., and F.D. Resources, M.A. Data curation, M.A., A.V., F.D., and M.G. Writing—original draft preparation, M.A., A.V., and F.E. Writing—review and editing, M.A. and M.G. Visualization, M.A. and L.P. Supervision, M.A. Project administration, M.A. Funding acquisition, M.A. All authors have read and agreed to the published version of the manuscript.

Funding: This research received no external funding.

Acknowledgments: The SAVEMEDCOASTS Project, funded by the EU (Agreement Number: ECHO/SUB/2016/742473/PREV16, coordinator Marco Anzidei) and Istituto Nazionale di Geofisica e Vulcanologia partially supported this study. We are thankful to Vincenzo Sepe of the INGV, for his contribution during preliminary GPS/RTK surveys.

Conflicts of Interest: The authors declare no conflict of interest.

References

1. Church, J.A.; Clark, P.U.; Cazenave, A.; Gregory, J.M.; Jevrejeva, S.; Levermann, A.; Merrifield, M.A.; Milne, G.A.; Nerem, R.S.; Nunn, P.D.; et al. Sea-level rise by 2100. *Science* **2013**, *342*, 1445. [CrossRef] [PubMed]

2. Kopp, R.E.; Kemp, A.C.; Bittermann, K.; Horton, B.P.; Donnelly, J.D.; Gehrels, W.R.; Hay, C.C.; Mitrovica, J.X.; Morrow, E.D.; Rahmstorf, S. Temperature-driven global sea-level variability in the Common Era. *Proc. Natl. Acad. Sci. USA* **2016**. [CrossRef] [PubMed]

3. Church, J.A.; Clark, P.U.; Cazenave, A.; Gregory, J.; Jevrejeva, S.; Levermann, A.; Merrifield, M.; Milne, G.; Nerem, R.S.; Nunn, P.; et al. Sea Level Change Supplementary Material. In *Climate Change 2013: The Physical Science Basis*; Contribution of Working Group I to the Fifth Assessment Report of the Intergovernmental Panel on Climate Change; Stocker, T.F., Qin, D., Plattner, G.-K., Tignor, M., Allen, S.K., Boschung, J., Nauels, A., Xia, Y., Bex, V., Midgley, P.M., Eds.; Cambridge University Press: Cambridge, UK; New York, NY, USA, 2013.

4. Bamber, J.L.; Oppenheimer, M.; Koppd, R.E.; Aspinallf, W.P.; Cooke, R.M. Ice Sheet Contributions to Future Sea-Level Rise from Structured Expert Judgment. 2019. Available online: www.pnas.org/cgi/doi/10.1073/pnas.1817205116 (accessed on 8 April 2019).

5. Veermer, M.; Rahmstorf, S. Global sea level linked to global temperature. *PNAS* **2009**, *106*, 21527–21532. [CrossRef] [PubMed]

6. Casella, E.; Rovere, A.; Pedroncini, A.; Mucerino, L.; Casella, M.; Cusati, L.A.; Vacchi, M.; Ferrari, M.; Firpo, M. Study of wave runup using numerical models and low altitude aerial photogrammetry: A tool for coastal management. *Estuar. Coast. Shelf Sci.* **2014**, *149*, 160–167. [CrossRef]

7. Antonioli, F.; Anzidei, M.; Amorosi, A.; Lo Presti, V.; Mastronuzzi, G.; Deiana, G.; De Falco, G.; Fontana, A.; Fontolan, G.; Lisco, S.; et al. Sea-level rise and potential drowning of the Italian coastal plains: Flooding risk scenarios for 2100. *Quat. Sci. Rev.* **2017**, *158*, 29–43. [CrossRef]

8. Lambeck, K.; Antonioli, F.; Anzidei, M.; Ferranti, L.; Leoni, G.; Scicchitano, G. and Silenzi, S. Sea level change along the Italian coast during the Holocene and projections for the future. *Quat. Int.* **2011**, *232*, 250–257. [CrossRef]

9. Ravanelli, R.; Riguzzi, F.; Anzidei, M.; Vecchio, A.; Nigro, L.; Spagnoli, F.; Crespi, M. Sea level rise scenario for 2100 AD for the archaeological site of Motya. *Rend. Lincei. Sci. Fis. e Nat.* **2019**. [CrossRef]

10. Anzidei, M.; Lambeck, K.; Antonioli, F.; Furlani, S.; Mastronuzzi, G.; Serpelloni, E.; Vannucci, G. Coastal structure, sea-level changes and vertical motion of the land in the Mediterranean. *Geol. Soc. Lond. Spec. Publ.* **2014**, *388*, 453–479. [CrossRef]

11. Colomina, I.; Molina, P. Unmanned aerial systems for photogrammetry and remote sensing: A review. *ISPRS J. Photogramm. Remote Sens.* **2014**, *92*, 79–97. [CrossRef]

12. Anzidei, M.; Bosman, A.; Carluccio, R.; Casalbore, D.; D'Ajello Caracciolo, F.; Esposito, A.; Nicolosi, I.; Pietrantonio, G.; Vecchio, A.; Carmisciano, C.; et al. Flooding scenarios due to land subsidence and sea-level rise: A case study for Lipari Island (Italy). *Terra Nova* **2017**, *29*, 44–51. [CrossRef]

13. Enei, F. *Santa Severa tra Leggenda e Realtà Storica. Pyrgi e il Castello di Santa Severa Alla Luce Delle Recenti Scoperte*; WordPress: Grotte di Castro, Municipality of Santa Marinella, Italy, 2013; pp. 1–413.

14. Colonna, G. *Il santuario di Pyrgi dalle origini mitistoriche agli altorilievi frontonali dei Sette e di Leucotea*; Scienze dell'Antichità; Poligrafico dello Stato: Roma, Italy, 2000.

15. Enei, F. *Pyrgi Sommersa. Ricognizioni archeologiche subacquee nel porto dell'antica Caere*; Museo civico Santa Marinella: Santa Marinella, Italy, 2008; pp. 1–115.

16. Enei, F. *Dal sito di Pyrgi, antico porto di Caere, nuovi dati per lo studio della linea di costa di epoca etrusca*; Serra, F., Ed.; Archaeologia Maritima Mediterranea: Pisa, Italy, 2013; pp. 165–176.

17. Mattei, M.; Conticelli, S.; Giordano, G. The Tyrrhenian margin geological setting: From the Apennine orogeny to the K-rich volcanism. In *The Colli Albani Volcano*; Special Publications of IAVCEI; Funiciello, R., Giordano, G., Eds.; Geological Society: London, UK, 2010; Volume 3, pp. 7–27.

18. Sestini, G.; Bruni, P.; Sagri, M. The Flysch Basins of the Northern Apennines: A Review of Facies and of Cretaceous-Neogene Evolution. *Memorie della Società Geologica Italiana* **1986**, *31*, 87–106.

19. Abbate, E.; Sagri, M. The eugeosynclinal sequences. *Sediment. Geol.* **1970**, *4*, 251–340. [CrossRef]

20. Alberti, A.; Bertini, M.; Del Bono, G.L.; Nappi, G.; Salvati, L. *Note illustrative della Carta Geologica d'Italia alla scala 1: 100000*; Tuscania-Civitavecchia Fogli 136–142; Poligrafica & Cartevalori: Ercolano, Italy, 1970.

21. Chiocchini, U.; Gisotti, G.; Macioce, A.; Manna, F.; Bolasco, A.; Lucarini, C.; Patrizi, G.M. Environmental geology problems in the Tyrrhenian coastal area of Santa Marinella, Province of Rome, central Italy. *Environ. Geol.* **1997**, *3*, 1–8. [CrossRef]

22. Ferranti, L.; Antonioli, F.; Mauz, B.; Amorosi, A.; Dai Pra, G.; Mastronuzzi, G.; Monaco, C.; Orrù, P.; Pappalardo, M.; Radtke, U.; et al. Markers of the last interglacial sea-level high stand along the coast of Italy: Tectonic implications. *Quat. Int.* **2006**, *145*, 30–54. [CrossRef]

23. Ferranti, L.; Antonioli, F.; Anzidei, M.; Monaco, C.; Stocchi, P. The timescale and spatial extent of vertical tectonic motions in Italy: Insights from coastal tectonic studies. *Rendiconti Online Societa Geologica Italiana* **2010**, *11*, 683–684.

24. Karner, D.; Marra, F.; Florindo, F.; Boschi, E. Pulsed uplift estimated from terrace elevations in the coast of Rome: Evidence for a new phase of volcanic activity? *Earth Planet. Sci. Lett.* **2001**, *188*, 135–148. [CrossRef]

25. Lambeck, K.; Anzidei, M.; Antonioli, F.; Benini, A.; Esposito, A. Sea level in Roman time in the Central Mediterranean and implications for recent change. *Earth Planet. Sci. Lett.* **2004**, *224*, 563–575. [CrossRef]

26. Avallone, A.; Selvaggi, G.; D'Anastasio, E.; D'Agostino, N.; Pietrantonio, G.; Riguzzi, F.; Serpelloni, E.; Anzidei, M.; Casula, G.; Cecere, G.; et al. The RING network: Improvement of a GPS velocity field in the central Mediterranean. *Ann. Geophys.* **2010**, *53*, 39–54.

27. Ullman, S. The interpretation of structure from motion. *Proc. R. Soc. Lond. Ser. B* **1979**, *203*, 405–426.

28. Sammari, C.; Koutitonsky, V.G.; Moussa, M. Sea level variability and tidal resonance in the Gulf of Gabes, Tunisia. *Cont. Shelf Res.* **2006**, *26*, 338–350. [CrossRef]

29. Lambeck, K.; Woodroffe, C.D.; Antonioli, F.; Anzidei, M.; Geherls, W.R.; Laborel, J.; Wright, A. Palaeoenvironmental records, geophysical modelling and reconstruction of sea-level trends and variability on centennial and longer time scales. In *Understanding Sea Level Rise and Variability*; Wiley-Blackwell: Oxford, UK, 2010; pp. 61–121.

30. Antonioli, F.; Lo Presti, V.; Anzidei, M.; Deiana, G.; de Sabata, E.; Ferranti, L.; Furlani, S.; Mastronuzzi, G.; Orru, P.E.; Pagliarulo, R.; et al. Tidal notches in Mediterranean Sea: A comprehensive analysis. *Quat. Sci. Rev.* **2015**, *119*, 66–84. [CrossRef]

31. Corchete, V. The high-resolution gravimetric geoid of Italy: ITG2009. *J. Afr. Earth Sci.* **2010**, *58*, 580–584. [CrossRef]

32. Serpelloni, E.; Faccenna, C.; Spada, G.; Dong, D.; Williams, S.D.P. Vertical GPS ground motion rates in the Euro-Mediterranean region: New evidence of velocity gradients at different spatial scales along the Nubia-Eurasia plate boundary. *J. Geophys. Res. Solid Earth* **2013**, *118*, 6003–6024. [CrossRef]

33. Serpelloni, E.; Pintori, F.; Gualandi, A.; Scoccimarro, E.; Cavaliere, A.; Anderlini, L.; Belardinelli, M.E.; Todesco, M. Hydrologically Induced Karst Deformation: Insights from GPS Measurements in the Adria-Eurasia Plate Boundary Zone. *J. Geophys. Res. Solid Earth* **2017**, *123*, 4413–4430. [CrossRef]

34. Herring, T.; King, R.W.; McClusky, S. *GAMIT Reference Manual, Release 10.4*; Massachussetts Institute of Technology: Cambridge, MA, USA, 2010.

35. Altamimi, Z.; Collilieux, X.; Métivier, L. ITRF2008: An improved solution of the international terrestrial reference frame. *J. Geod.* **2011**, *85*, 457. [CrossRef]

36. Devoti, R.; D'Agostino, N.; Serpelloni, E.; Pietrantonio, G.; Riguzzi, F.; Avallone, A.; Cavaliere, A.; Cheloni, D.; Cecere, G.; D'Ambrosio, C.; et al. The mediterranean crustal motion map compiled at INGV. *Ann Geophys.* **2017**. [CrossRef]

37. Guidoboni, E.; Comastri, A.; Traina, G. *Catalogue of Ancient Earthquakes in the Mediterranean Area up to the 10th Century*; Istituto Nazionale di Geofisica: Roma, Italy, 1994.

38. Rovere, A.; Antonioli, F.; Enei, F.; Giorgi, S. Relative sea level change at the archaeological site of Pyrgi (Santa Severa, Rome) during the last seven millennia. *Quat. Int.* **2011**, *232*, 82–91. [CrossRef]

39. Boccotti, P. *Wave Mechanics for Ocean Engineering*; Elsevier Science: Amsterdam, The Netherlands, 2000; Volume 64.

40. Arena, F.; Laface, V.; Malara, G.; Romolo, A.; Viviano, A.; Fiamma, V.; Sannino, G.; Carillo, A. Wave climate analysis for the design of wave energy harvesters in the Mediterranean Sea. *Renew. Energy* **2015**, *77*, 125–141. [CrossRef]

41. Weggel, J.R. Maximum breaker height. *J. Waterw. Harb. Coast. Eng. Div.* **1972**, *98*, 529–548.

42. Komar, P.D.; Gaughan, M.K. Airy wave theory and breaker height prediction. In Proceedings of the 13th Conference on Coastal Engineering, Vancouver, BC, Canada, 29 January 1972. [CrossRef]

43. U.S. Army Corps of Engineers. *Coastal Engineering Manual, Part II*; U.S. Army Corps of Engineers: Washington, DC, USA, 2006.

44. Mase, H. Random wave run-up height on gentle slope. *J. Waterw. Port Coast. Ocean Eng.* **1989**, *115*, 649–661. [CrossRef]

45. Lambeck, K.; Anzidei, M.; Antonioli, F.; Benini, A.; Verrubbi, V. Tyrrhenian sea level at 2000 BP: Evidence from Roman age fish tanks and their geological calibration. *Rend. Fis. Acc. Lincei* **2018**, *29*, 69–80. [CrossRef]

46. Anzidei, M.; Scicchitano, G.; Tarascio, S.; De Guidi, G.; Monaco, C.; Barreca, G.; Mazza, G.; Serpelloni, E.; Vecchio, A. Coastal retreat and marine flooding scenario for 2100: A case study along the coast of Maddalena Peninsula (southeastern Sicily). *Geogr. Fis. Dinam. Quat.* **2018**, *41*, 5–16. [CrossRef]

47. Vecchio, A.; Anzidei, M.; Serpelloni, E.; Florindo, F. Natural Variability and Vertical Land Motion Contributions in the Mediterranean Sea-Level Records over the Last Two Centuries and Projections for 2100. *Water* **2019**, *11*, 148. [CrossRef]

48. Reimann, L.; Vafeidis, A.T.; Brown, S.; Hinkel, J.; Tol, R.S.J. Mediterranean UNESCO World Heritage at risk from coastal flooding and erosion due to sea-level rise. *Nat. Comm.* **2018**, *9*, 4161. [CrossRef]

49. Masselink, G.; Hughes, M.G. *Introduction to Coastal Processes and Geomorphology*; McCann, S.B., Ed.; Edward Arnold: London, UK, 1980; p. 354.

50. Sobel, A.H.; Camargo, S.J.; Hall, T.M.; Lee, C.Y.; Tippett, M.K.; Wing, A.A. Human influence on tropical cyclone intensity. *Science* **2016**, *353*, 242–246. [CrossRef]

51. Rahmstorf, S. Rising hazard of storm-surge flooding. *PNAS* **2017**, *114*, 45. [CrossRef]

52. Menéndez, M.; Woodworth, P. Changes in extreme high water levels based on a quasi-global tide-gauge data set. *J. Geophys. Res.* **2010**, *115*, C10011. [CrossRef]

53. Lowe, J.A.; Gregory, J.M. The effects of climate change on storm surges around the United Kingdom. *Philos. Trans. R. Soc.* **2005**, *363*. [CrossRef]

Journal of
Marine Science and Engineering

Article

UAV Photogrammetry and Ground Surveys as a Mapping Tool for Quickly Monitoring Shoreline and Beach Changes

Antonio Zanutta *, Alessandro Lambertini and Luca Vittuari

Department of Civil, Chemical, Environmental and Materials Engineering (DICAM), University of Bologna, Viale Risorgimento 2, 40136 Bologna, Italy; alessandro.lambertini@unibo.it (A.L.); luca.vittuari@unibo.it (L.V.)
* Correspondence: antonio.zanutta@unibo.it; Tel.: +39-051-2093111

Received: 20 December 2019; Accepted: 14 January 2020; Published: 18 January 2020

Abstract: The aim of this work is to evaluate UAV photogrammetric and GNSS techniques to investigate coastal zone morphological changes due to both natural and anthropogenic factors. Monitoring morphological beach change and coastline evolution trends is necessary to plan efficient maintenance work, sand refill and engineering structures to avoid coastal drift. The test area is located on the Northern Adriatic coast, a few kilometres from Ravenna (Italy). Three multi-temporal UAV surveys were performed using UAVs supported by GCPs, and Post Processed Kinematic (PPK) surveys were carried out to produce three-dimensional models to be used for comparison and validation. The statistical method based on Crossover Error Analysis was used to assess the empirical accuracy of the PPK surveys. GNSS surveys were then adopted to evaluate the accuracy of the 2019 photogrammetric DTMs. A multi-temporal analysis was carried out by gathering LiDAR dataset (2013) provided by the "Ministero dell'Ambiente e della Tutela del Territorio e del Mare" (MATTM), 1:5000 Regional Technical Cartography (CTR, 1998; DBTR 2013), and 1:5000 AGEA orthophotos (2008, 2011). The digitization of shoreline position on multi-temporal orthophotos and maps, together with DTM comparison, permitted historical coastal changes to be highlighted.

Keywords: coastline mapping; geomatics; SfM photogrammetry; network RTK

---

1. Introduction

The morphology of coastal areas is influenced by both natural phenomena and anthropic activities. The most characteristic feature of a coastal area is its shoreline, the boundary between land and a water surface.

Its location varies continuously over time due to different factors which modify the shape of the beach [1], which can create huge economic problems for the tourist industry. Coastal erosion has always existed and, over the centuries, has contributed to shaping coastal areas, through the delicate environmental balance between fluvial regulation and rising sea levels. The increase of anthropic activity along the coasts has damaged the equilibrium, accentuating the dynamics of erosive—progradation.

The coastal erosion affecting the entire North Adriatic coast puts at risks local infrastructure, environment and the tourist economy. Italian Regional Authorities, have adopted an "Integrated Coastal Zone Management Plan" involving strategies based on coastal protection structures, submerged breakwaters and, as preferred strategy for coastal protection, beach replenishment [2–4].

Over the years, research on strategies for coastal preservation has been carried out. In this context the role of geomatic techniques is fundamental for mapping and monitoring [5].

Existing shoreline mapping techniques are characterized by different accuracy, expense and training time requirements.

Topographic observations and remote sensing methods are the most commonly used techniques to detect the position of the shoreline, and monitor the shape and extension of coastal areas [6]. Point-based measurements are carried out by means of different measuring procedures with Levels, Total Stations (TS) and Global Navigation Satellite System (GNSS) instruments.

GNSS surveys (by static, fast static, kinematic, real time kinematic methods) and those carried out by means of TS (triangulation, trilateration, intersection—resection, traverses, radial sides shots methods and more generally 3D or 2D networks) together with geometric or trigonometric levelling require physical contact with the portion of land to be measured. The main limitation of these methods is the huge amount of time required and the technical difficulties involved in surveying.

The geomatic surveying methods, such as aero photogrammetric surveys [7,8], unmanned aerial vehicle (UAV) [9–15], remote sensing with satellite images [12,16], airborne light detection and ranging (LiDAR) [17–22], synthetic aperture radar (SAR) [23,24], video systems [25–29] are widespread techniques to measure wide areas, generate digital terrain models (DTMs) and maps, and to quantify coastal changes by multi-temporal comparisons.

The shoreline is constantly changing therefore all the abovementioned remote sensing techniques are fundamental because allow to acquire synchronous information that can be analysed in laboratory at a later stage. To verify the efficacy of the beach nourishment works, monitoring topo-bathimetric surveys are necessary.

The nourishment methods based on the artificial transport of sand in the areas subject to erosion, can be implemented by land, transporting the sand by truck or taking it from submarine borrowing areas located offshore and transporting the sand in situ by pipes. Given the high cost of handling the sand transport, the monitoring interventions must allow the best resolution, with the ease of use method and an excellent quality/cost ratio.

Due to the high mobility of the sand surface, for natural causes or following human intervention (e.g., for the protection of the beach touristic infrastructures during the winter), the level of significance of the comparison between two surveys of the beach surface can be quantified in ±10 cm vertically. This precision threshold can be achieved, quickly and economically by using UAVs combined with image matching photogrammetric procedures [30–33] over small-medium sized areas, making this combination an efficient tool for high resolution investigation for coastal management.

GNSS provides high accuracy coordinates of points and can be used to describe surface and volumetric changes. GNSS and photogrammetry are complementary because GNSS is fundamental for solving image orientation procedures and assessing accuracy, while photogrammetry can produce dense DSMs with radiometric content.

The aim of this work is to evaluate photogrammetric and GNSS techniques for performing 3D surveys of coastal environments in the context of coastline change studies. The techniques used have the advantage of carrying out an almost synchronous survey on medium-sized areas with the required resolution. Three multi-temporal UAV surveys were performed using UAVs. In order to allow a robust possibility of a quantitative comparison between the surfaces surveyed in subsequent epochs, the positions of the ground control points (GCPs), were surveyed, by means of a network real time kinematic (NRTK) service. In this way, in fact, the connection to the international geodetic reference system is guaranteed by a network of permanent GNSS stations. At the same time, post-processed kinematic (PPK) surveys were carried out to produce three-dimensional models by means of interpolations of the tracks used for comparison and validation. DTMs, sections and orthophotos were obtained through photogrammetry processing with similar geometric resolution. Vectorial elaborates were numerically compared to understand and quantify the geomorphological changes while orthophotos were used to highlight variations in the coastal shoreline.

A test was carried out to evaluate the effects of the number and distribution of GCPs used to orient the blocks of images acquired from the UAV. A subset of GCPs arranged along the same route was adopted and the differences between the DTMs obtained were analysed. Furthermore, multi-temporal analysis was carried out by collecting a LiDAR dataset (CC BY-SA 3.0 IT) surveyed in

2013 provided by "Ministero dell'Ambiente e della Tutela del Territorio e del Mare" (MATTM) within the context of "Piano Straordinario di Telerilevamento Ambientale" (PST-A), 1:5000 Regional Technical Cartography (CTR, 1998) Regional Topographic Database (DBTR 2013), and 1:5000 AGEA orthophotos (2008, 2011) from the GeoER, GIS Office of Emilia Romagna Region, Italy ("Archivio Cartografico Regione Emilia—Romagna", https://geoportale.regione.emilia-romagna.it/it).

2. Study Area

The area surveyed is located between the villages of Punta Marina and Lido Adriano, on the Northern Adriatic coast 7 km away from the city of Ravenna, Italy (Figure 1). The southern jetty of Ravenna is 7 km to the North. The nearest river is Fiumi Uniti, which flows 3 km south of the area.

Figure 1. The survey area (highlighted in the red box), is located on the northern Adriatic coast, close to Lido Adriano, Emilia Romagna Region (Italy), 7 km from the southern jetty of Ravenna (**A**). The four cross-sections analysed (S1–S4) are identified by dashed red lines. The Regional Technical Cartography (2013) shown in black is superimposed on the AGEA Orthophoto (2011; EPSG:25833). The old shoreline (CTR 1998) is represented with black crosses (**B**).

The area studied is approximately 450 m long in the NW-SE direction and 110 m wide. It is a portion of beach especially chosen to test several geomatic techniques of surveying. Four cross-sections labelled S1–S4 from North to South (Figure 1) are defined in the study area for the analysis carried out in this work. S1–S4 are perpendicular to the average shoreline and equally spaced with a distance of 100 m between them.

The area is of considerable touristic interest, as the beach is an important resource for recreation, and also for research into the engineering of coastal zone management. From a geological point of view, the beach is characterised by fine sands covering muddy-clayey materials from alluvial deposits and ancient swamps.

The extension of the beach is a function of river and marine coastal processes influenced by anthropic activities and natural phenomena. The predominantly erosive tendency of the area is due to the limited supply of river sediments and is amplified by the rate of subsidence [26,34]. The position of the coastline is also influenced by tides and winds. As in all of the northern Adriatic, the tides are strongly asymmetrical, with a diurnal and semi-diurnal component. The characteristic tidal sea-level variations at Lido Adriano beach are 30–40 cm at neap tide and 80–90 cm at spring tide. The tidal effect combined with storm surges, and the action of winds can produce significant morphological changes to the beach [2].

The main winds are Bora and Scirocco [35]. The Bora is a cold, intense wind that comes from the NE and is more common in winter. The Scirocco is a warm wind that usually comes from the SE and is typical of the spring and autumn seasons and is one of the main factors responsible for events associated with high water levels. The beach is protected by offshore coastal protection structures that minimize erosive processes. To the North and South of the study area there are detached breakwaters perpendicular to the coast while offshore, semi-submerged barriers are parallel to the coast protecting the beach.

3. Data Availability

The current dataset of observations was acquired in three steps (2017, 2018, 2019) by means of DJI Matrice 600 (M600) and Spark UAVs (Figure 2, Table 1) in order to produce DTMs and orthophotos.

Figure 2. Panoramic image of the surveyed area and DJI M600 during the flight (**A**); UAVs used for the surveys: DJI Spark (**B**), detail of DJI M600 (**C**).

Table 1. Summary of the photogrammetric dataset characteristics.

Data	07/12/2017	25/05/2018	24/05/2019
Camera Model	M600_X5	M600_X5	FC1102
Resolution (pixel)	4000 × 2250	4000 × 2250	3968 × 2976
Sensor dimension (mm)	17.3 × 13.0	17.3 × 13.0	6.17 × 4.56
Focal Length (mm)	15	15	4.49
Pixel Size (µm)	4.33	4.33	1.57
N° images	450	99	296
Relative height (m)	113 *	66.7	53.1
Mean scale	7533	4447	11826
Ground resolution cm/pixel	3.26	1.92	1.86

* Maximum flight height of a set of experimental flights on the same area.

These UAVs are multi-rotor systems:

DJI Matrice 600, 525 × 480 × 640 mm, max take-off weight 15.5 kg, with a max. payload of 5.5 kg and flight time of approximately 30 min, equipped with a DJI M600-X5 camera;
DJI Spark, 143 × 143 × 55 mm, lightened to a weight less than 300g and flight time of approximately 16 min, equipped with a DJI FC 1102 camera.

The on-board digital cameras provide useful imagery for aerial photogrammetric surveys and are georeferenced with an onboard GNSS receiver, which provides the cameras with positional accuracy of about 10 m. The three multi-temporal UAV flights (2017, 2018, 2019) were drawn up using an orthophoto of the area and the relative height of flight was selected to obtain a ground sampling distance (GSD) of about 2 cm (Table 1). Longitudinal and side image overlapping within photogrammetric blocks were fixed respectively at 80% and 70%. The direction of the flight strips was established within the flight planning software in order to minimize the number of images and the time of flight. The positions of GCPs were surveyed by means of NRTK technique using 50 cm diameter plywood targets, surveyed before each fly. These GCPs were visible in the images and were adopted to orient the blocks of images as described in the following paragraph. PPK surveys were carried out, for a direct survey of crossing profiles of the beach during the 2019 photogrammetric survey, for DTM accuracy evaluations (Figure 3).

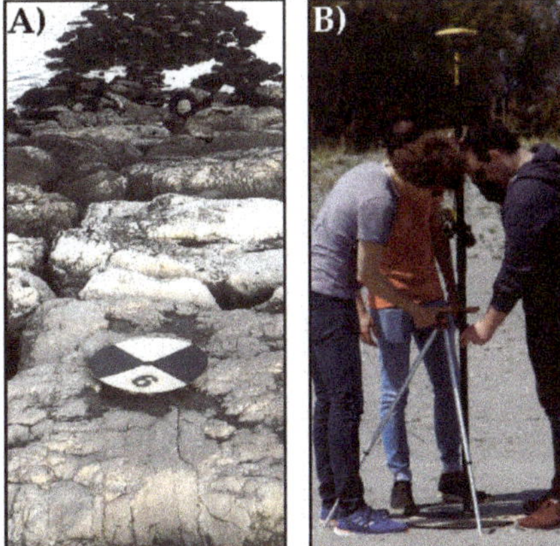

Figure 3. A numbered GCP in the survey area, made with high-density heavy plywood (**A**); GCP survey by means of NRTK technique (**B**); PPK survey performed by mounting the surveying system in a rigid backpack (**C**).

The accuracies in positioning of RTK, NRTK and PPK methods, in our test conditions, are in the order of a couple of cm horizontally and 3–5 cm vertically. In order to carry out a multi-temporal analysis, DTM, maps and orthophotos were collected from the "Ministero dell'Ambiente e della Tutela del Territorio e del Mare" (MATTM), and from the GeoER, GIS Office of Emilia Romagna Region, (Italy; Table 2).

Table 2. Summary of the available data used for multitemporal comparison.

Source	Date	Origin Description	Scale	Accuracy (m)
CTR	1977–1998	Scan of Traditional map	1:5000	2
DBTR	1997–2013	Cartographic representation of the contents of the DBTR	1:5000	2
AGEA	2008	Orthophotos; GSD 0.5 m; GCPs from CTR 1:5000	1:10000	4
AGEA	2011	Orthophotos; GSD 0.5 m; GCPs from CTR 1:5000	1:10000	4
LiDAR	2013	Digital Surface Model; GSD 2.0 m		0.3

CTR—Regional Technical Map; DBTR—Regional Topographic Database; AGEA—"AGenzia per le Erogazioni in Agricoltura"; GSD—Ground Sampling Distance (m/pixel); GCP—Ground Control Point.

4. Photogrammetric Analysis

Dense point clouds and orthophotos were generated with a commercial software (Agisoft Metashape, Professional Edition 1.5.5, Agisoft LLC, St. Petersburg, Russia) which performs automatic tie point extraction and feature matching with bundle block adjustment [30]. The software is based on the structure from motion (SfM) algorithms [30–33]. In the first stage an algorithm is applied that detects points in the source photos which are stable for viewpoint and lighting variation. Secondly, a descriptor for each point based on its local neighbourhood is generated. These descriptors are used to detect correspondence between the photos. Intrinsic and extrinsic orientation parameters of the camera are than solved and refined by bundle-adjustment algorithm.

A self-calibration procedure is used in the image orientation process adopting a unique camera model for each project, evaluating the interior orientation parameters: Focal length, coordinates of principal point, lens distortion (K1, K2, K3, p1, p2), affinity and shear parameters. GNSS-assisted block orientation was applied using as *a priori* values the GNSS coordinates of the camera projection centre (PC) locations, stored in the ancillary information of the images contained in the Exchangeable image file format (Exif) file. A dense surface reconstruction is produced from the aligned images using processing methods based on pairwise depth map computation.

Finally, after a mesh calculation, texture mapping is applied and then source photos are projected onto the model (Supplementary Materials, Figures S1 and S2). GCPs were adopted to frame the bundle block adjustment and to register DTMs in the same datum and mapping projection (EPSG: 25833; Table 3). The orientation process was based on a set of GCPs uniformly distributed over the study area (Figure 4).

Table 3. Summary of the Orientation process and DTM characteristics.

Data	07/12/2017	25/05/2018	24/05/2019
GCP [1] n°	9	12	9
RMSE [2] (cm)	15	6.3	2.3
DTM [3] resolution (cm)	2.7	4.3	3.3
DTM [3] density (points/m^2)	1360	532	903

[1] Ground Control Point; [2] Root Mean Square Error; [3] Digital Terrain Model.

The accuracy of the aerial triangulation process is summarized in Table 3. The last survey was identified as the most accurate by root mean square errors. The distribution of GCPs is decisive and at the same time critical in coastal contexts. Uncontrolled deformations in the derived models can be induced when GCPs in the study area are not distributed homogeneously, as may happen in long stretches of the coast. In our experience it was possible to evaluate this effect, by processing the dataset surveyed in 2019 a second time with only four GCPs aligned with each other. This processing introduced a rotation of the model as shown in Figure 4.

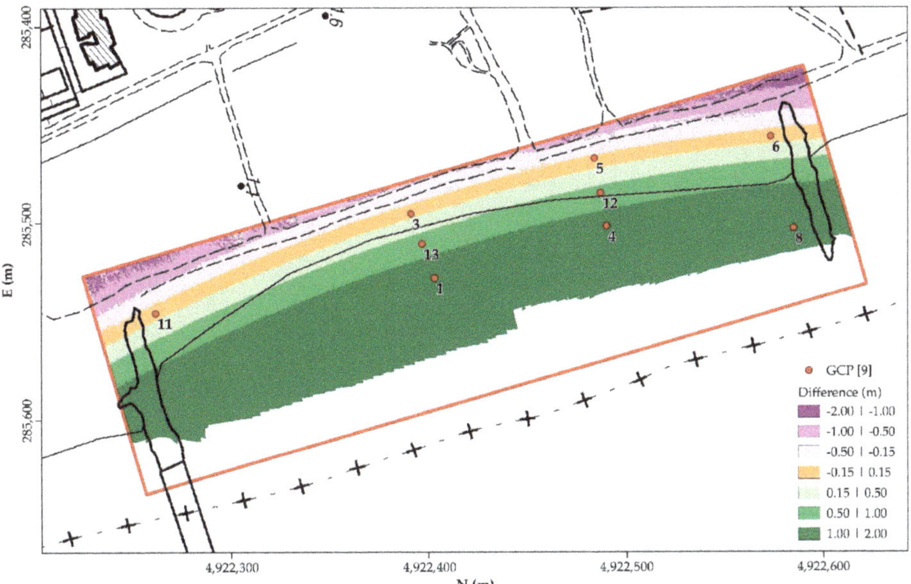

Figure 4. The vertical difference in 2019 dataset between the reference survey processed with all 9 GCPs and the test with only 4 aligned GCPs (ID: 11, 3, 5, 6). Axis of rotation highlighted in orange. The Regional Technical Cartography (2013) is shown in black.

5. PPK and NRTK GNSS Surveys

In order to allow georeferencing of the photogrammetric blocks within the UTM-ETRS89 grid system, nine GCPs were measured before the acquisition of the images, with a Topcon Hiper V Series GNSS receiver. NRTK mode using a correction mode based on virtual reference stations (VRS) through the service Topcon, Netgeo (http://www.netgeo.it/) was adopted.

Easily transportable GCPs were constructed from of 50 cm diameter plywood targets, printed with a black and white pattern and a code for easy detection. They were distributed uniformly over the area to be surveyed before flights.

The number of GCPs was chosen to minimize survey time and therefore costs. At the same time the total number of GCPs was significantly higher than the minimum necessary. This allowed a more rigorous control of the study area and redundancy in the photogrammetric orientation procedures using some of the surveyed coordinates as check points. The expected precision is 2–3 cm in horizontal coordinates and 5–10 cm for elevation, also considering the precision of the geoid model used to transform the ellipsoidal heights into orthometric ones (Geoid model ITALGEO 2005) [36].

The NRTK stationing over GCPs was fixed at two minutes of static acquisition for each point. PPK surveys were carried out to produce three-dimensional models of the topographical surface, by mounting the surveying system in rigid backpacks (GPS rover), enabling it to track the 3D profile followed by the operators.

A reference GNSS station was established close to the beach, to collect GNSS data at 1s sync rate, throughout each survey. For georeferencing of PPK surveys, the coordinates of the master station were estimated in the ETRF2000 system through the classical static differenced approach using the simultaneous acquisitions carried out at the closest EUREF reference station MSEL (Medicina, Bologna).

Then the processing of the kinematic data was performed with the open source software RTKLIB v 2.4.3 (www.rtklib.com). Considering the small extension of the studied area (a few hundred metres) absolute errors of a few centimetres characterised the survey. For the PPK tracks, the positions of the GNSS rover stations were evaluated in kinematic mode starting from the known master station coordinates, obtaining a single position for each acquisition time (1 s).

Latitudes and longitudes were then projected in the EPSG 25833, using ITALGEO 2005 geoid model to transform the ellipsoidal heights into orthometric ones. The sparse points were then interpolated using a linear Kriging approach.

6. Empirical Accuracy Assessment of PPK and Photogrammetric Surveys

In order to check the internal accuracy of the PPK surveys carried out during the 2019 survey, the crossover error analysis (CEA) was performed on the tracks surveyed, according to Hsu [37]. The procedure is based on height determination and comparison at the intersections of PPK track lines [38].

A routine was elaborated by means of QGIS software to obtain precise height values along the track intersections. The procedure used was articulated in several steps. Firstly, each 3D point of the track was connected by lines ("linestring" function) and transformed in a Well-Known Binary (WKB) format. Then, by means of line Intersection function, each line crossover was detected. For each planimetric intersecting PPK line, a height interpolation was performed at crossover point. Finally, height differences were computed at crossover points.

As result, the histogram of the observed residuals exhibits an almost normal distribution with a mean value close to zero and a RMSE of 0.03 m for 697 intersection points (Figure 5).

The dense static clouds of points recorded during operator stops in the tracks were eliminated manually in order to optimize comparison. Statistical results (Figure 5C) reveal that these data are consistent with the expected accuracy for kinematic surveys in walking mode with antenna mounted on backpacks and so are sufficiently accurate to validate photogrammetric DTMs.

Then the accuracy of the photogrammetric survey was evaluated firstly by comparing the GCP coordinates extracted manually from DTMs with those from NRTK surveys as reference. The statistical results are shown in Table 4.

Table 4. Statistics of the differences between the coordinates of the GCPs and those measured on the photogrammetric DTMs.

Data	07/12/2017	25/05/2018	24/05/2019
Mean (m)	−0.01	−0.01	0.00
RMSE (m)	0.09	0.10	0.02
Max (m)	0.24	0.83	0.06
Min (m)	0.06	0.03	0.01

These values are affected both by mismatching errors and the local density of DTMs.

PPK GNSS surveys provided the second opportunity to evaluate the accuracy of the 2019 photogrammetric DTM. The assessment was carried out both by computing the height difference of each PPK point with respect to photogrammetric derived DTM, and calculating the difference between two DTMs: The first derived by interpolating PPK points and the second obtained by photogrammetric processing (Figure 6).

In particular, the first comparison was made to minimise the effects of interpolation, as points located along the tracks were directly compared with the corresponding ones located on photogrammetric DTM, while the second test was made by interpolating PPK points with a kriging algorithm, then transforming them in raster format with cell size of 10 cm. Statistical data and the range of differences shown in Figure 6 confirm good agreement between the models. In fact, both approaches provided similar results.

Statistical values of cases A and B (Figure 6) are similar because the area surveyed is predominantly flat, so as a result the spacing between the PPK tracks was sufficient to represent the shape of the beach.

Shorelines were digitized on orthophotos and technical regional maps (Emilia Romagna Region, CTR and DBTR 1:5000 scale) and analysed from a qualitative and quantitative point of view.

The photogrammetric model of 2019 was then compared with LiDAR DTM and the differences evaluated numerically.

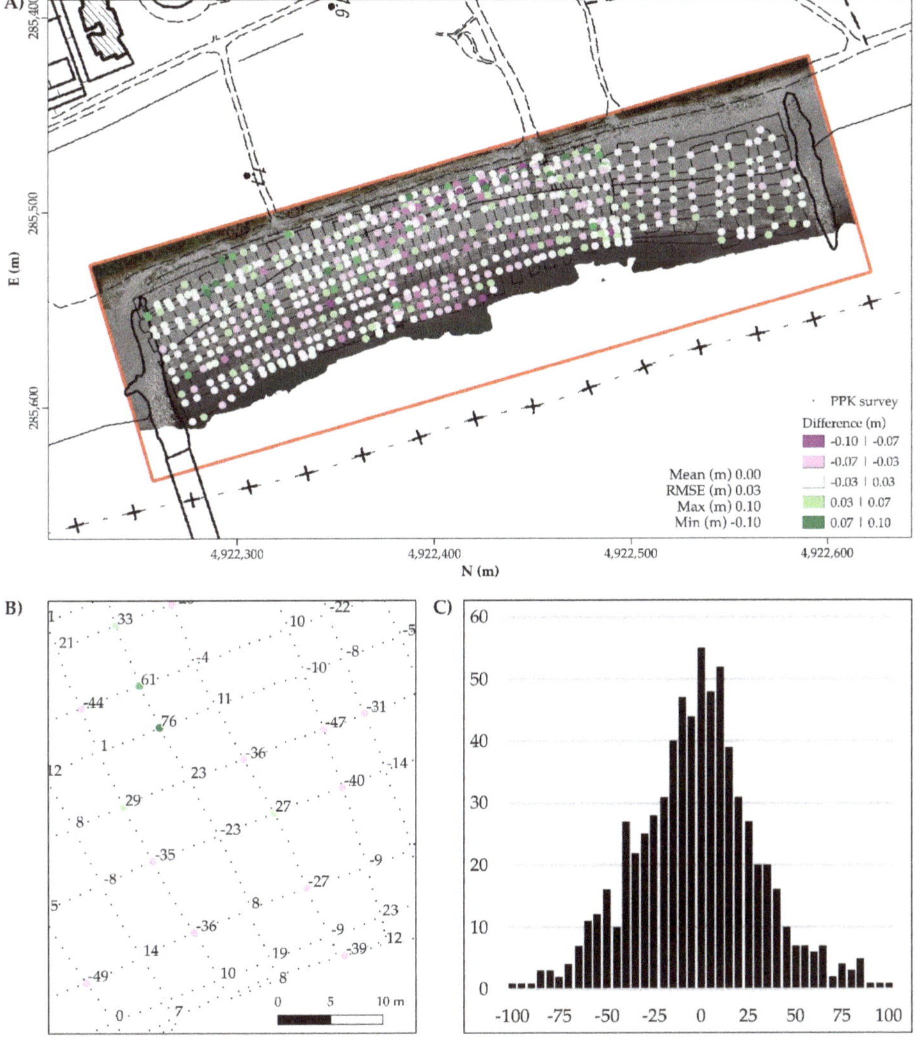

Figure 5. Results of the Crossover Error Analysis of track intersections of the 2019 survey. (**A**) map of the 2019 PPK tracks showing the height residuals in different colours, M mean value, RMSE Root Mean Square Error; (**B**) detail of a portion of the surveyed profiles with differences at the crossing points in mm (the same colours of Figure 5A); (**C**) Histogram distribution of the height differences expressed in mm.

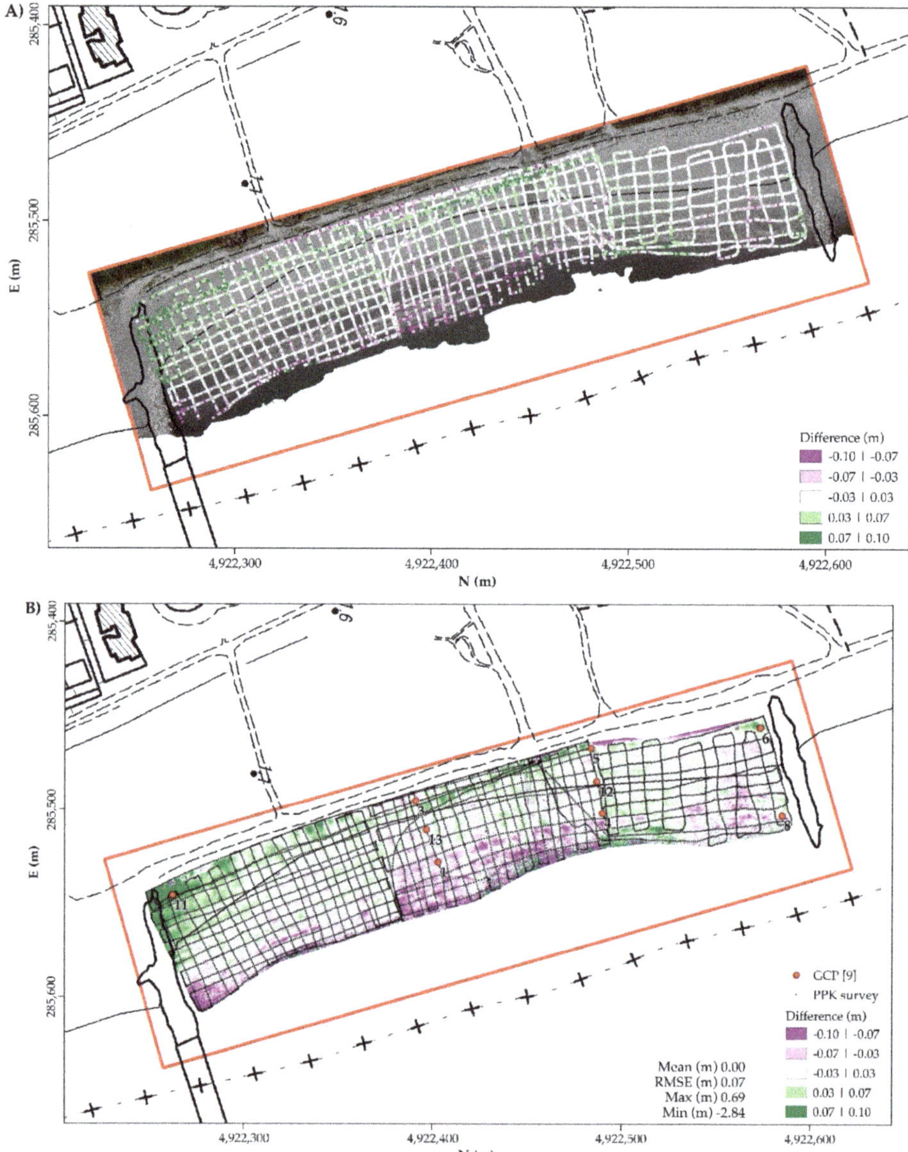

Figure 6. Map of the differences between the PPK and photogrammetric DTM. (**A**) height difference over PPK points; (**B**) NRTK assisted- photogrammetric DTMs comparison. The red box highlights the survey area. The GCP location is shown by the red dots.

7. Results

The shoreline of the studied area, located at the village of Lido Adriano in Ravenna province is subjected to natural phenomena and anthropic actions which have led to the construction of groins to protect the beach [39,40]. Both shoreline and sand surface are constantly changing. UAV combined to GNSS techniques may be an efficient tool for coastal management. Surveys carried out by means of these techniques are synchronous, characterised by the necessary accuracy and quickly to perform.

The digitization of shoreline position on multi-temporal orthophotos and maps, together with DTM comparison, permitted historical coastal changes to be highlighted (Figures 7 and 8). Distances between shorelines were determined along the S1–S4 transects using the most recent measurement as reference, shown in Table 5.

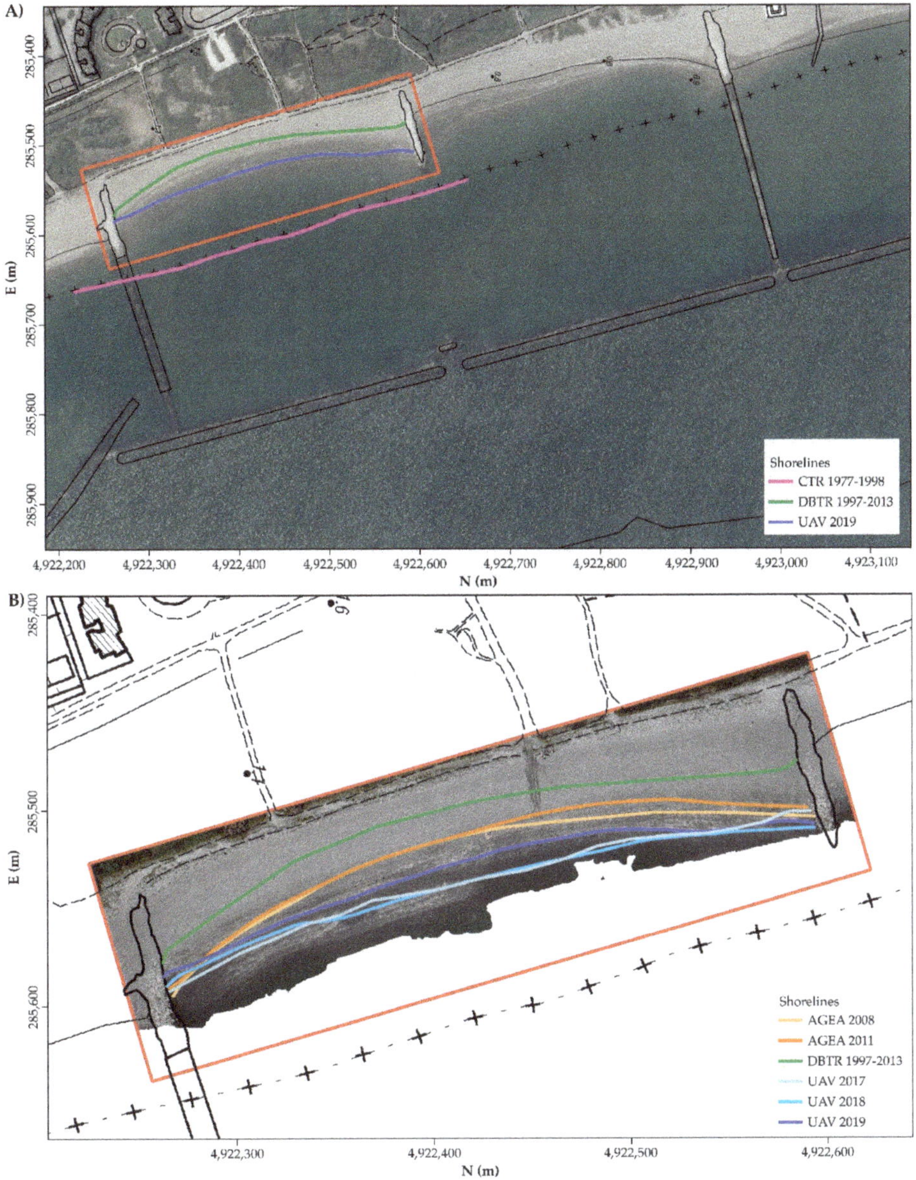

Figure 7. Multi-temporal shorelines: (**A**) Key map of the layers Orthophoto AGEA 2011 and DBTR 2013 showing the studied area; (**B**) Detail image of the UAV derived Orthophoto 2019 and the DBTR 2013.

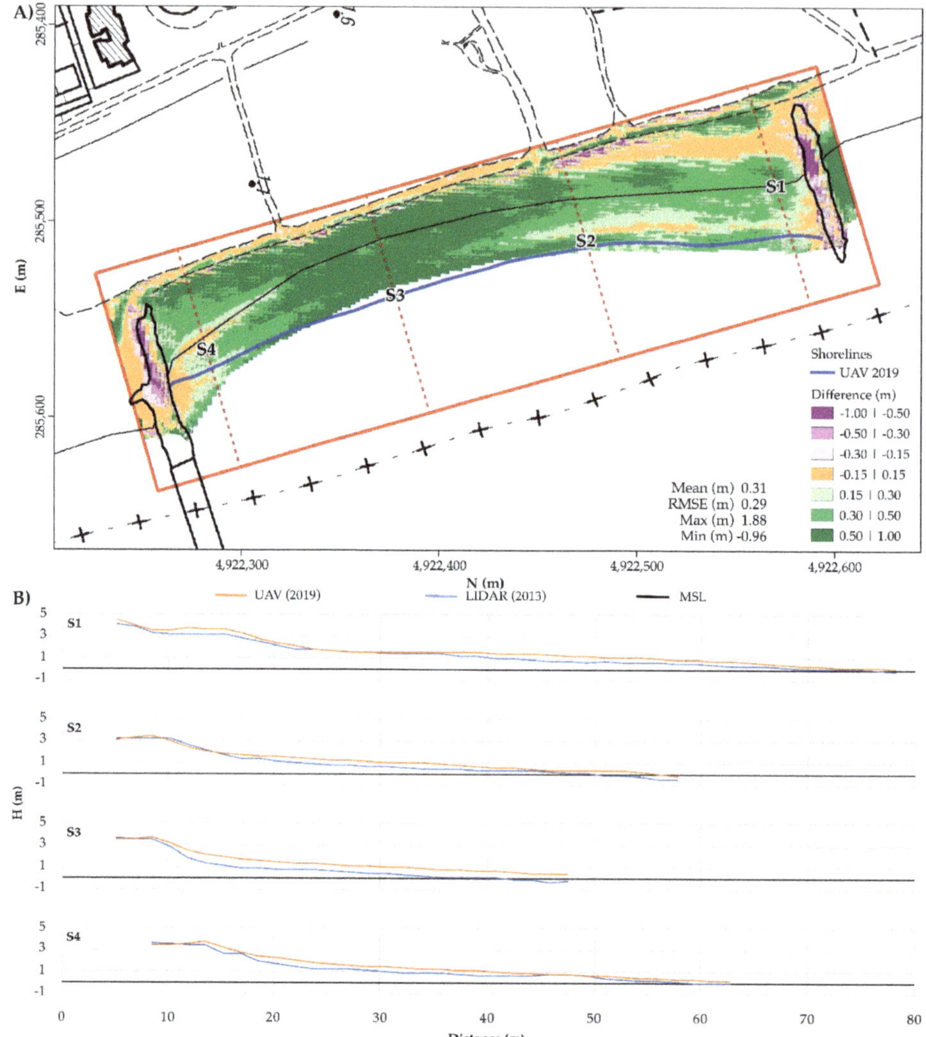

Figure 8. Multi-temporal comparison and transect analysis in Lido Adriano area: (**A**) difference between 2019 UAV derived DTM and 2013 LIDAR DTM; (**B**) cross-shore profiles of the beach (S1–S4), 2019 UAV derived DTM (red) and 2013 LIDAR DTM (blue).

Maximum recession between shorelines is recorded comparing CTR to DBTR. AGEA 2008 and 2011 shorelines remain essentially stable in the central and southern parts while a small difference is detected towards the North. Accretion trends are observed matching these shorelines with the UAV 2019 survey. A small erosional effect can be observed in the studied area from 2017 to 2019 (Figure 7).

In Figure 8A the differences obtained by comparing the DTM detected with 2019 UAV with the 2013 LiDAR model are shown in different colours. The variations in the central part of the beach show the seasonally man-made actions carried out to flatten and enlarge the beach.

The positive values (green), clearly show the operations of beach nourishment. Not significant changes can be seen along the dune parallel to the shoreline (orange). Negative values on the breakwaters perpendicular to the shoreline can be observed in the period considered. In the 2019 model these blocks of rocks were lower, with different areal extension and show a rate of settling of a few cm per year. For these comparisons high vegetation was removed from the UAV derived model.

Table 5. Differences expressed in m between the multitemporal shorelines along the four cross-sections (S1–S4) and statistical data. The most recent shoreline detected with the 2019 UAV survey, was used as reference. Values are in metres. CTR—Regional Technical Map; DBTR—Regional Topographic Database, AGEA—"AGenzia per le Erogazioni in Agricoltura".

Source	CTR	DBTR	AGEA	AGEA
Date	1977–1998	1977–2013	2008	2011
S1	−53.77	26.56	3.36	7.98
S2	−76.29	25.69	7.81	14.32
S3	−81.11	29.83	13.51	13.56
S4	−71.23	15.98	−0.30	0.49
Mean (m)	−70.60	24.52	6.10	9.09
Std Dev (m)	11.92	5.96	5.95	6.39
Max (m)	−53.77	29.83	13.51	14.32
Min (m)	−81.11	15.98	−0.30	0.49

8. Discussion and Conclusions

Beach morphology and long-term shoreline changes derive from the sum of natural phenomena and anthropic activities. The coastal shoreline is the indicator traditionally used to define the trend of the sandy coasts, highlighting depositional or erosional phenomena. In beaches subject to anthropic activities like those of the Northern Adriatic coast, this interpretation becomes invalid, because shorelines migrate landward or seaward depending not only on changing sea-level or subsidence of coastal regions, but also on the presence of low-crested structures and nourishment carried out periodically for beach protection (Arpae Emilia-Romagna, www.arpae.it).

In order to use beaches for tourism it is fundamental to preserve their width. Therefore, monitoring morphological beach changes and coastline evolution trends is necessary to plan efficient maintenance work as well as replenishment and constructing engineering structures to avoid coastal drift.

This study highlights that changes in the shoreline derived from orthophoto and regional maps do not follow a constant positive or negative rate because they are due to human intervention designed to protect the coast. Since the 1980s a number of defence structures have been built and replenishment repeatedly carried out. The artificial supply of sand, and the periodical flattening of the beach by bulldozer have contributed to decreasing erosional rates and widened the beach [2,41,42].

The purpose of this study was to test fast and cheap geomatic techniques for coastline mapping and detecting shoreline changes. Three UAV photogrammetric surveys integrated with GNSS measurement of GCPs and PPK tracks on a beach protected by a system of breakwaters located in Lido Adriano, Italy was presented.

The images acquired by UAV were elaborated with image matching algorithms [30–33] that allowed maps to be produced automatically and DTM to be extracted with high spatial resolution and accuracy.

In addition to investigating the shape of the beach, quantitative and qualitative analyses of the differences were carried out using freely available multitemporal maps and DTMs from the "Ministero dell'Ambiente e della Tutela del Territorio e del Mare", GeoER, GIS Office of Emilia Romagna Region, and "AGenzia per le Erogazioni in Agricoltura".

The survey procedures adopted were cheaper and faster compared to the traditional ones. The kinematic positioning techniques PPK, RTK and NRTK allow, in undisturbed operating conditions, better accuracies than 5 cm in planimetry and height. Applying these surveying techniques in the estimation of the positions of the shutter centres of the cameras, together with a cross-strip flight, and just a single GCP positioned in the central area of the block, it is possible to obtain accuracies comparable to those obtainable through the use of a homogeneous distribution of GCPs [43]. In case of significant offsets between the GNSS antenna and the position of the camera, the presence of a synchronized onboard strapdown inertial navigation system (SINS) mechanically connected to the camera, allows to correct the orientation up to 0.10°–0.05°, even if the camera is mounted within a gimbal.

Low-cost, professional, commercial UAVs can be used with good results to produce maps, detect topographical changes, and estimate variations in rate and volume. This represents a tool for coastal managers and the regional authorities to improve their "Integrated Coastal Zone Management Plans". In further developments, the use of NRTK GNSS receivers onboard of the UAVs and high-performance inertial platforms directly connected and synced to the UAV cameras, may reduce the number of GCPs in the survey of long shorelines. Nowadays long-distance surveys are limited by the flight autonomy, which is significantly longer for fixed wing UAVs and the evolving flight regulations.

Supplementary Materials: The following are available online at http://www.mdpi.com/2077-1312/8/1/52/s1, Figure S1: 2019 UAV derived Orthophoto, Figure S2: DTM, EPSG:25833, GSD:0.05m.

Author Contributions: Conceptualization of the study, L.V., A.Z.; methodology, A.L., L.V., A.Z.; software, A.L.; validation, L.V.; investigation, L.V., A.L., A.Z.; data curation, A.L., L.V., A.Z.; writing—original draft, A.L., L.V., A.Z.; writing—review and editing, A.L., L.V., A.Z.; supervision, L.V. All authors have read and agreed to the published version of the manuscript.

Funding: This study received no external funding.

Acknowledgments: We are thankful to Ing. Luca Poluzzi, Ing. Luca Tavasci for their valuable help during the fieldwork activities. Special thanks to the student Ing. Riccardo Collina for his help in the GNSS-assisted block orientation test to produce Figure 4, in the frame of his second level Civil Engineering thesis (University of Bologna). The authors would like to thank MATTM, Emilia Romagna Region and AGEA for providing LiDAR data, maps and multitemporal orthophotos. The authors thank the Guest editor Donatella Dominici, the assistant editor Zara Liu, and the four referees for their constructive comments and suggestions, which greatly improved the manuscript.

Conflicts of Interest: The authors declare no conflict of interest.

References

1. Smith, A.W.S.; Jackson, L.A. The variability in width of the visible beach. *Shore Beach* **1992**, *60*, 7–15.
2. Armaroli, C.; Ciavola, P.; Perini, L.; Lorito, S.; Valentini, A.; Masina, M. Critical storm thresholds for significant morphological changes and damage along the Emilia-Romagna coastline, Italy. *Geomorphology* **2012**, *143–144*, 34–51. [CrossRef]
3. Cantasano, N.; Pellicone, G.; Ietto, F. Integrated coastal zone management in Italy: A gap between science and policy. *J. Coast Conserv.* **2017**, *21*, 317–325. [CrossRef]
4. Ciavola, P.; Armaroli, C.; Chiggiato, J.; Valentini, A.; Deserti, M.; Perini, L.; Luciani, P. Impact of storms along the coastline of Emilia-Romagna: The morphological signature on the Ravenna coastline (Italy). *J. Coast. Res.* **2007**, 540–544.
5. Moore, L.J. Shoreline mapping techniques. *J. Coast. Res.* **2000**, *16*, 111–124. [CrossRef]
6. Boak, E.; Turner, I.L. Shoreline definition and detection: A review. *J. Coast. Res.* **2005**, *21*, 688–703. [CrossRef]
7. Michalowska, K.; Glowienka, E.; Pekala, A. Spatial-Temporal detection of changes on the southern coast of the Baltic sea based on multitemporal aerial photographs. In Proceedings of the International Archives of the Photogrammetry, Remote Sensing and Spatial Information Sciences—ISPRS Archives, Prague, Czech Republic, 12–19 July 2016; Volume 41, pp. 49–53. [CrossRef]
8. Schwarzer, K.; Diesing, M.; Larson, M.; Niedermeyer, R.O.; Schumacher, W.; Furmanczyk, K. Coastline evolution at different time scales—Examples from the Pomeranian Bight, southern Baltic Sea. *Mar. Geol.* **2003**, *194*, 79–101. [CrossRef]
9. Benassai, G.; Aucelli, P.; Budillon, G.; De Stefano, M.; Di Luccio, D.; Di Paola, G.; Montella, R.; Mucerino, L.; Sica, M.; Pennetta, M. Rip current evidence by hydrodynamic simulations, bathymetric surveys and UAV observation. *Nat. Hazards Earth Syst. Sci.* **2017**, *17*, 1493–1503. [CrossRef]

10. Casella, E.; Rovere, A.; Pedroncini, A.; Stark, C.P.; Casella, M.; Ferrari, M.; Firpo, M. Drones as tools for monitoring beach topography changes in the Ligurian Sea (NW Mediterranean). *Geo-Mar. Lett.* **2016**, *36*, 151–163. [CrossRef]

11. Duo, E.; Trembanis, A.C.; Dohner, S.; Grottoli, E.; Ciavola, P. Local-scale post-event assessments with GPS and UAV based quick-response surveys: A pilot case from the Emilia—Romagna (Italy). *Coast. Nat. Hazards Earth Syst. Sci.* **2018**, *18*, 2969–2989. [CrossRef]

12. Giordan, D.; Notti, D.; Villa, A.; Zucca, F.; Calò, F.; Pepe, A.; Dutto, F.; Pari, P.; Baldo, M.; Allasia, P. Low cost, multiscale and multi-sensor application for flooded area mapping. *Nat. Hazards Earth Syst. Sci.* **2018**, *18*, 1493–1516. [CrossRef]

13. Mancini, F.; Dubbini, M.; Gattelli, M.; Stecchi, F.; Fabbri, S.; Gabbianelli, G. Using Unmanned Aerial Vehicles (UAV) for High-Resolution Reconstruction of Topography: The Structure from Motion Approach on Coastal Environments. *Remote Sens.* **2013**, *5*, 6880–6898. [CrossRef]

14. Turner, I.L.; Harley, M.D.; Drummond, C.D. UAVs for coastal surveying. *Coast. Eng.* **2016**, *114*, 19–24. [CrossRef]

15. Van Puijenbroek, M.E.; Nolet, B.; de Groot, C.; Suomalainen, A.V.; Riksen, J.M.; Berendse, M.J.P.M.; Limpens, F.J. Exploring the contributions of vegetation and dune size to early dune development using unmanned aerial vehicle (UAV) imaging. *Biogeosciences* **2017**, *14*, 5533–5549. [CrossRef]

16. Guariglia, A.; Buonamassa, A.; Losurdo, A.; Saladino, R.; Trivigno, M.L.; Zaccagnino, A.; Colangelo, A. A multisource approach for coastline mapping and identification of shoreline changes. *Ann. Geophys.* **2006**, *49*, 295–304. [CrossRef]

17. Bazzichetto, M.; Malavasi, M.; Acosta, A.T.R.; Carranza, M.L. How does dune morphology shape coastal EC habitats occurrence? A remote sensing approach using airborne LiDAR on the Mediterranean coast. *Ecol. Indic.* **2016**, *71*, 618–626. [CrossRef]

18. Charlton, M.E.; Large, A.R.G.; Fuller, I.C. Application of airbourne LiDAR in river environments: The River Coquet, Northumberland, UK. *Earth Surf. Process. Landf.* **2003**, *28*, 299–306. [CrossRef]

19. Le Mauff, B.; Juigner, M.; Ba, A.; Robin, M.; Launeau, P.; Fattal, P. Coastal monitoring solutions of the geomorphological response of beach-dune systems using multi-temporal LiDAR datasets (Vendée coast, France). *Geomorphology* **2018**, *304*, 121–140. [CrossRef]

20. Middleton, J.H.; Cooke, C.G.; Kearney, E.T.; Mumford, P.J.; Mole, M.A.; Nippard, G.J.; Rizos, C.; Splinter, K.D.; Turner, I.L. Resolution and accuracy of an airborne scanning laser system for beach surveys. *J. Atmos. Ocean. Technol.* **2013**, *30*, 2452–2464. [CrossRef]

21. Pye, K.; Blott, S.J. Assessment of beach and dune erosion and accretion using LiDAR: Impact of the stormy 2013-14 winter and longer term trends on the Sefton Coast, UK. *Geomorphology* **2016**, *266*, 146–167. [CrossRef]

22. Stockdonf, H.F.; Sallenger, A.H., Jr.; List, J.H.; Holman, R.A. Estimation of shoreline position and change using airborne topographic lidar data. *J. Coast. Res.* **2002**, *18*, 502–513.

23. Nunziata, F.; Buono, A.; Migliaccio, M.; Benassai, G. Dual-polarimetric C-and X-band SAR data for coastline extraction. *IEEE J. Sel. Top. Appl. Earth Obs. Remote Sens.* **2016**, *9*, 4921–4928. [CrossRef]

24. Tajima, Y.; Wu, L.; Fuse, T.; Shimozono, T.; Sato, S. Study on shoreline monitoring system based on satellite SAR imagery. *Coast. Eng. J.* **2019**, *61*, 401–421. [CrossRef]

25. Alesheikh, A.A.; Ghorbanali, A.; Nouri, N. Coastline change detection using remote sensing. *Int. J. Environ. Sci. Technol.* **2007**, *4*, 61–66. [CrossRef]

26. Archetti, R. Quantifying the evolution of a beach protected by low crested structures using video monitoring. *J. Coast. Res.* **2009**, *25*, 884–899. [CrossRef]

27. Brignone, M.; Schiaffino, C.F.; Isla, F.I.; Ferrari, M. A system for beach video-monitoring: Beachkeeper plus. *Comput. Geosci.* **2012**, *49*, 53–61. [CrossRef]

28. Holman, R.A.; Stanley, J. The history and technical capabilities of Argus. *Coast. Eng.* **2007**, *54*, 477–491. [CrossRef]

29. Kroon, A.; Aarninkhof, S.G.J.; Archetti, R.; Armaroli, C.; Gonzalez, M.; Medri, S.; Osorio, A.; Aagaard, T.; Davidson, M.A.; Holman, R.A.; et al. Application of remote sensing video systems for coastline management problems. *Coast. Eng.* **2007**, *54*, 493–505. [CrossRef]

30. Triggs, B.; McLauchlan, P.; Hartley, R.; Fitzgibbon, A. Bundle Adjustment-A Modern Synthesis. In Proceedings of the International Workshop on Vision Algorithms, ICCV'99, Corfu, Greece, 20–25 September 1999; Springer: Berlin/Heidelberg, Germany, 2000; pp. 298–372. [CrossRef]

31. Seitz, S.; Curless, B.; Diebel, J.; Scharstein, D.; Szeliski, R. A Comparison and Evaluation of Multi-View Stereo Reconstruction Algorithms. In Proceedings of the 2006 IEEE Computer Society Conference on Computer Vision and Pattern Recognition (CVPR 2006), New York, NY, USA, 17–22 June 2006; pp. 519–528. [CrossRef]

32. Snavely, N.; Seitz, S.M.; Szeliski, R. Photo tourism: Exploring photo collections in 3D. *ACM Trans. Graph.* **2006**, *25*, 835–846. [CrossRef]

33. Ullman, S. The interpretation of structure from motion. *Proc. R. Soc. Lond. B* **1979**, *203*, 405–426.

34. Teatini, P.; Ferronato, M.; Gambolati, G.; Bertoni, W.; Gonella, M. A century of land subsidence in Ravenna, Italy. *Environ. Geol.* **2005**, *47*, 831–846. [CrossRef]

35. Zavatarelli, M.; Pinardi, N. The Adriatic Sea modelling system: A nested approach. *Ann. Geophys.* **2003**, *21*, 345–364. [CrossRef]

36. Barzaghi, R.; Borghi, A.; Carrion, D.; Sona, G. Refining the estimate of the Italian quasi-geoid. *Boll. Geod. Sci. Affin.* **2007**, *66*, 145–160.

37. Hsu, S.K. XCORR: A cross-over technique to adjust track data. *Comput. Geosci.* **1995**, *21*, 259–271. [CrossRef]

38. Baldi, P.; Bonvalot, S.; Briole, P.; Marsella, M. Digital photogrammetry and kinematic GPS applied to the monitoring of Vulcano Island, Aeolian Arc, Italy. *Geophys. J. Int.* **2000**, *142*, 801–811. [CrossRef]

39. Giambastiani, B.M.; Greggio, N.; Sistilli, F.; Fabbri, S.; Scarelli, F.; Candiago, S.; Gabbianelli, G. RIGED-RA project-Restoration and management of Coastal Dunes in the Northern Adriatic Coast, Ravenna Area-Italy. In Proceedings of the IOP Conference Series Earth Environment Science, World Multidisciplinary Earth Sciences Symposium (WMESS 2016), Prague, Czech Republic, 5–9 September 2016; Volume 44, pp. 38–52. [CrossRef]

40. Sytnik, O.; Del Río, L.; Greggio, N.; Bonetti, J. Historical shoreline trend analysis and drivers of coastal change along the Ravenna coast, NE Adriatic. *Environ. Earth Sci.* **2018**, *77*, 779. [CrossRef]

41. Harley, M.D.; Ciavola, P. Managing local coastal inundation risk using real-time forecasts and artificial dune placements. *Coast. Eng.* **2013**, *77*, 77–90. [CrossRef]

42. Scarelli, F.M.; Sistilli, F.; Fabbri, S.; Cantelli, L.; Barboza, E.; Gabbianelli, G. Seasonal dune and beach monitoring using photogrammetry from UAV surveys to apply in the ICZM on the Ravenna coast (Emilia-Romagna, Italy). *Remote Sens. Appl. Soc. Environ.* **2017**, *7*, 27–39. [CrossRef]

43. Forlani, G.; Dall'Asta, E.; Diotri, F.; Morra di Cella, U.; Roncella, R.; Marina Santise, M. Quality Assessment of DSMs Produced from UAV Flights Georeferenced with On-Board RTK Positioning. *Remote Sens.* **2018**, *10*, 311. [CrossRef]

Journal of
Marine Science and Engineering

Article

Shoreline Extraction Based on an Active Connection Matrix (ACM) Image Enhancement Strategy

Sara Zollini [1,*,†], Maria Alicandro [1,†], María Cuevas-González [2,†], Valerio Baiocchi [3,†], Donatella Dominici [1,†] and Paolo Massimo Buscema [4,5,†]

1   DICEAA, Department of Civil, Environmental Engineering and Architecture, Via Gronchi 18, 67100 L'Aquila, Italy; maria.alicandro@univaq.it (M.A.); donatella.dominici@univaq.it (D.D.)
2   Centre Tecnològic de Telecomunicacions de Catalunya (CTTC), Division of Geomatics, Av. Gauss 7, E-08860 Castelldefels, Barcelona, Spain; maria.cuevas@cttc.es
3   DICEA, Sapienza University of Rome, via Eudossiana, 18, 00184 Roma, Italy; valerio.baiocchi@uniroma1.it
4   Semeion Research Center of Sciences of Communication, Via Sersale 117, 00128 Rome, Italy; m.buscema@semeion.it
5   Department of Mathematical and Statistical Sciences, University of Colorado, Denver, CO 82017, USA
*   Correspondence: sara.zollini@graduate.univaq.it; Tel.:+39-0862-434116
†   These authors contributed equally to this work.

Received: 19 November 2019; Accepted: 17 December 2019; Published: 23 December 2019

Abstract: Coastal environments are facing constant changes over time due to their dynamic nature and geological, geomorphological, hydrodynamic, biological, climatic and anthropogenic factors. For these reasons, the monitoring of these areas is crucial for the safeguarding of the cultural heritage and the populations living there. The focus of this paper is shoreline extraction by means of an experimental algorithm, called J-Net Dynamic (Semeion Research Center of Sciences of Communication, Rome, Italy). It was tested on two types of image: a very high resolution (VHR) multispectral image (WorldView-2) and a high resolution (HR) radar synthetic aperture radar (SAR) image (Sentinel-1). The extracted shorelines were compared with those manually digitized for both images independently. The results obtained with the J-Net Dynamic algorithm were also compared with common algorithms, widely used in the literature, including the WorldView water index and the Canny edge detector. The results show that the experimental algorithm is more effective than the others, as it improves shoreline extraction accuracy both in the optical and SAR images.

Keywords: remote sensing; satellite images; synthetic aperture radar (SAR); Sentinel-1; WorldView-2; shoreline extraction; coastline extraction; active connection matrix (ACM); J-Net Dynamic; edge detection; canny edge detector

---

1. Introduction

The coastal environment is an extraordinary natural resource, not only from the point of view of the cultural heritage but for hosting resources that can be measured in terms of economic assets. It is a dynamic environment, subject to continuous and constant transformation. The coastal area is indeed a highly dynamic system where erosion and deposition phenomena are influenced by various factors. According to [1], the factors responsible for changes in coastal areas may be grouped into geological and geomorphological, hydrodynamic, biological, climatic and anthropogenic factors. The coastal environment, where large percentages of the global population live, change rapidly due to its dynamic nature. For this reason, the availability of up-to-date information on its state is of great interest.

The theoretical definition of "shoreline" is merely the transition between the sea and the land [2,3], and in general, between the land surface and the surface of a water body, such as an ocean, sea or lake. The coastal zone is, instead, that area of land and water that borders the shoreline and extends landward [4]. In [5–7] other definitions about coastal areas can be found. In theory, the concept is very simple and intuitive but its application is actually a complex task because of the temporal variability of the shoreline itself. The temporal variability develops on scales profoundly different from instantaneous to secular variations, and it depends on various factors, including wave motion, tides and winds, but also erosion and deposition. For this reason, the results obtained from surveys on the field or remote sensing are generally indicators of the actual shoreline.

According to [7], there are three main subdivisions of shoreline indicators:

- Characteristics visible by an operator on an aerial or remote sensing image;
- Intersection between a tidal datum and a digital terrain model or a coastal profile;
- Characteristics of multispectral images identified by automatic algorithms not necessarily visible to an unaided operator.

In this work, we will experiment with several indicators of the third type by comparing them with each other and with those of the first type. It is important to underline, however, that those things which are extracted are indicators of the shoreline and not the shoreline itself.

Several survey methodologies are used for the detection of shorelines: from traditional topographical and GNSS (Global Navigation Satellite System) surveys to remote sensing techniques (aerial and unmanned aerial vehicle (UAV) photogrammetry, video systems and satellite optical and SAR images [8,9]), as widely discussed in another recent work [10]. In the last few years, the use of optical and SAR satellite images for automatic and semi-automatic shoreline extraction has complemented the traditional approach based on surveying methods and air-photo processing [11]. Several approaches have been proposed to outline the shoreline using remote sensing, such as edge detection, segmentation and classification approaches [6].

An edge in an image is a boundary where there is a change in some physical parameters and it is like a contour between two different regions. Many approaches have been used for edge extraction; for example, the Canny edge detector [12,13] or the snake model [14]. Segmentation methods can be divided into three categories: threshold-based, region-based and edge-based [15]. Concerning classification, they can be pixel-oriented, based on individual pixel classifications or object-oriented—which instead are able to group characteristics by aggregation in similar regions or polygons. Among the pixel-oriented classification, spectral indices have been successfully used to extract shorelines. In particular, water indices (WIs) and vegetation indices (VIs) have been developed, which are obtained by combining two or more spectral bands. Although VIs are mainly used for vegetation analysis, they have also been tested for shoreline mapping purposes [10,16]. WIs have been specifically developed for water body detection [17].

In this work, an experimental algorithm, called J-Net Dynamic [18,19], was tested both on a very high resolution (VHR) WorldView-2 optical image, and for the first time, on a high resolution (HR) Sentinel-1 SAR image. The J-Net Dynamic algorithm can be considered an edge operator, as it highlights the edges of the image and is part of the active connection matrix (ACM) system, which is a new unsupervised artificial adaptive system developed by Semeion Research Institute. The system can automatically extract features of the images—edges, tissue differentiation, etc., when they are activated by original non-linear equations. ACM activation reduces the image noise while maintaining the spatial resolution of high contrast structures. The J-Net Dynamic exploits the dynamic connections between a central pixel and its neighboring pixels. The relationships among these latter pixels with their neighbors is also taken into account, so the new image is created considering the totality of these contributions at every elaboration cycle. Every single neighbor pixel participates in the evolution of the ACM, which is finally used to create the pixel matrix with the new image. A detailed description of the algorithm can be found in Appendix A.

2. Study Area and Data Set

The Abruzzo region (Italy) coastline has a length of 125 km, of which 26 km is high coast and 99 km is sandy coast; the latter, therefore, accounts for 80% of the entire coastline and more than 50% is under erosion. The northern sector, between the Tronto River and Ortona, is characterized by low coasts; proceeding southwards, the coast is formed, up to Vasto, by mainly high coasts, but, at Vasto Marina and San Salvo the coasts are low and sandy. The average width of the Abruzzo beaches is often less than 100 m, and in some cases, it does not reach 50 m. The dominant winds are the Mistral and the Tramontana. Studies performed on wave measurements demonstrated that among the states of significant sea, that is wave heights greater than 0.5 m, the most frequent waves have heights greater than two meters and the most intense wave motions have heights between 3.5 and 6 m, characterized by an occurrence frequency of less than 5% [20].

The study of this paper was conducted on the sandy coast of Ortona (Abruzzo Region) in which three erosion phenomena have been reported on the coast, four active landslides that have effects on the coast and three landslide crags on the sea [21]. For this reason, according to the guidelines for the defense of the coast against erosion and the effects of climate change [22], one of the fundamental elements of a coastal information system is the extraction of the shoreline. The potential of the J-Net Dynamic algorithm for identifying the shoreline, in automatic or semiautomatic mode, from remote sensing, has been tested on two types of images: multispectral WorldView-2 and SAR Sentinel-1 satellite images. The Digital Globe WorldView-2 sensor, launched in October 2009, was the first VHR, 8-band, multispectral commercial satellite [23] with four standard (red, green, blue and near-infrared) and four new (coastal, yellow, red edge and near-infrared) spectral bands [24]. The images are provided with a panchromatic band (0.5 m) and eight multispectral bands (2 m) [25]. Moreover, depending on the level of processing, the images can be supplied as "Basic," "Standard," "Ortho Ready Standard Imagery" or "Stereo Pair" products [26]. A WorldView-2 image acquired on 29 June 2010 was analyzed in this work. This image was projected to UTM WGS84 reference system (WGS84/UTM Zone 33 N, EPSG: 32633). After pre-processing, described in Section 3.1, a pan-sharpening pre-treatment was also performed to merge the high geometric resolution of the panchromatic image with the multispectral bands. A Sentinel-1 SAR image has also been analyzed in this work. The Sentinel-1 mission is the European Radar Observatory for the Copernicus joint initiative of the European Commission (EC) and the European Space Agency (ESA) [27]. Sentinel-1 images are freely available and are also supplied with a level of processing that guarantees immediate use: level-0 provides raw images, level-1 Single Look Complex (SLC), level-1 Ground Range Detected (GRD), level-2 Ocean. Seven Sentinel-1 Interferometric Wide (IW) Level-1 (GRD) images (see Table 1) have been downloaded from Copernicus Open Access Hub website [28].

Table 1. List of the Sentinel-1 images used in this analysis.

Sentinel-1 Images	
Number	Date dd/mm/yyyy
1	12/01/2017
2	06/01/2017
3	31/12/2016
4	25/12/2016
5	19/12/2016
6	13/12/2016
7	07/12/2016

They consist of focused SAR data that have been detected, multi-looked and projected to ground range using the Earth ellipsoid model WGS84. The ellipsoid projection of the GRD products is corrected using the terrain height specified in the product general annotation. The terrain height varies in azimuth but is constant in range (but can be different for each IW sub-swath) [29]. The IW swath acquires data with 250 km swath at 5 m by 20 m spatial resolution. IW captures three sub-swaths using Terrain Observation with Progressive Scans SAR (TOPSAR) [30]. Ground range coordinates are the slant range coordinates projected onto the ellipsoid of the Earth. Pixel values represent detected amplitude. There is no information about the phase. The resulting product has approximately square resolution pixels and square pixel spacing with reduced speckle but reduced spatial resolution. For the IW GRD products, multi-looking is performed on each burst individually. All bursts in all sub-swaths are then seamlessly merged to form a single, contiguous, ground range detected image per polarization. Seven Sentinel-1 images are processed in order to better remove the speckle, although just the one in the middle was processed for shoreline extraction.

Information regarding tides was also taken into account to compare the instantaneous shoreline [31]. In Ortona area, the maximum high tide recorded in the tide tables is 0.6 m and the minimum height is –0.2 m, referenced to mean lower low water (MLLW). As the wave action on the Mediterranean coast is limited [32] the images can be considered at the same tidal moment.

3. Methodology

3.1. Shoreline Extraction with WorldView-2 Image

The WorldView-2 images are provided as Ortho Ready Standard products, so a pre-processing is required to correct the geometric distortions [16]. An orthorectification was performed using ERDAS Imagine 2015 [33], and rational polynomial coefficients (RPCs) were used to correct the geometric distortions. In addition, 17 ground control points (GCPs) and 8 check points (CPs) were used to improve the process [10]. Finally, the original spatial resolution of the multi-spectral bands was improved with the panchromatic one by applying the HCS (hyperspherical color dpace) resolution merge algorithm, designed for WorldView-2 [34] and implemented in Erdas Imagine. After the pre-processing, two algorithms were applied to extract the shoreline edges: the WVWI (WorldView Water Index) and the J-Net Dynamic. The WVWI (1) is a WI adjusted for the WorldView-2 images [35] and defined as:

$$WVWI = \frac{CB - NIR2}{CB + NIR2},$$ (1)

where CB and $NIR2$ are referred to the Coastal band (400–450 nm) and Near Infrared 2 band (860–900 nm), respectively. WVWI combines the Coastal and the Near-Infrared 2 bands pixel by pixel and it ranges mathematically between -1 and 1. WVWI tends to 1 in the presence of water and to 0 where sand is present. To vectorize the shoreline extracted from WVWI, a threshold was set in correspondence of the inflection point evaluated on some WVWI profiles, as it was determined in [36] and in a previous work [10] (Figure 1). The peaks, showed around 150 m on the profile, correspond to the rocks but, for the purpose of this paper, they were not taken into account.

A reclassification based on the threshold was performed and finally the shoreline was digitized with the Arcscan tools in ArcMap (ESRI, Redlands, California). Finally, a smoothing algorithm was used to improve the quality of vector line [37].

The J-Net Dynamic algorithm has been tested on a single band or band combinations. For the sake of brevity, in this work only the analysis of the NIR2 band, which gave the best results, is reported. The NIR2 band is indeed suitable for discriminating water from wet and dry sand, as it is almost completely absorbed by the water and possible perturbations are minimized by the shallow bottoms [38,39].

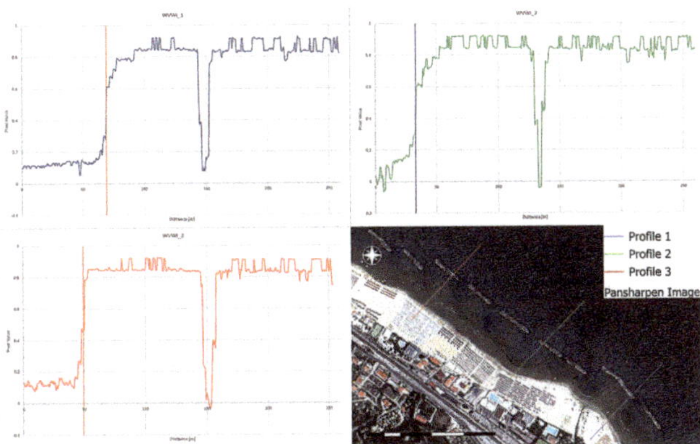

Figure 1. Spatial profiles of WVWI (WorldView Water Index): along the x-axis the distances (m) and along the y-axis the WVWI pixel values; the vertical line represents the inflection point in which the threshold between water and sand has been set.

The J-Net Dynamic algorithm, thoroughly described in the Appendix A, permits one to obtain a clear separation threshold between water and land (Figure 2).

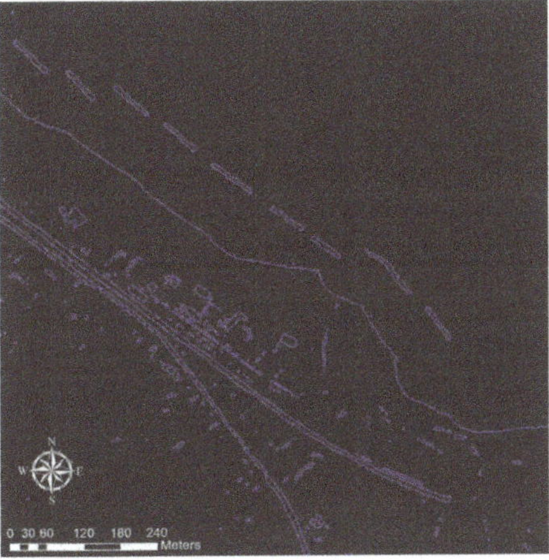

Figure 2. J-Net algorithm tested on NIR2 band of the WorldView-2 image.

The vectorization of the shoreline extracted by J-Net is performed in the same way of the WVWI.

3.2. Shoreline Extraction with Sentinel-1 Image

The pre-processing of Sentinel-1 images is a key issue of SAR elaboration because, for a better identification of the shoreline, it is essential to improve the image quality, and in this case, to reduce noise while preserving the edges. Pre-processing methods can be grouped into noise reduction and image correction [6]. SAR images are affected by speckle, an effect caused by the coherent radiation used by radar systems. It is like a salt and pepper effect that happens because each resolution cell associated with an extended target contains several scattering centers whose elementary returns, by positive or negative interference, originate light or dark image brightness [40]. There are some adaptive filters, such as the median filter, which are suitable for SAR speckle removal. The median filter uses the median values within a moving kernel in place of each pixel of the image. That way, it smooths the image without smoothing the edges [6]. The SNAP (Sentinel Application Platform) software is used to pre-process the Sentinel-1 SAR images. The pre-processing steps that were carried out are:

1. Thermal noise removal;
2. Apply Orbit filer;
3. Calibration to Beta0;
4. Coregistration;
5. Multitemporal de-speckle;
6. Range doppler terrain correction.

The pre-processing aims to enhance the image for a better interpretation. In particular, the purpose is to increase the contrast while preserving the edges, using bands, algorithms and polarizations which maximize the difference between coast and water. The following options were tested:

1. Use decibel bands or not;
2. Use polarization VV or VH;
3. Choose the filter for the multitemporal de-speckle.

Two masks, one on the sea and one on the mainland, have been created and statistics have been computed using the ENVI software (Harris Geospatial Solutions, Broomfield, Colorado, United States). For each study, the digital number (DN) minimum, maximum, mean and standard deviation are calculated. The results show that the contrast between sea and land is higher using the decibel bands and VH polarization. Among the filters computed with SNAP to reduce the speckle, the one which maximizes the contrast is the IDAN, but it presents the problem of excessively smoothing the edges. For this reason, the Lee filter is used, because it is a good compromise between keeping the spatial resolution and preserving the edges [41–43].

Finally, the image selected for analysis and shoreline extraction was the one in decibels, with the VH polarization and whose speckle had been reduced using the Lee filter. Figure 3 shows the image acquired on December 31, 2016 after the pre-processing.

Figure 3. Pre-processed and georeferenced Sentinel-1 image. Date: December 31, 2016. Units: decibels. Polarization: VH polarization. De-speckle filter: Lee.

After image pre-processing, edge detection algorithms are used to extract the shoreline. It is also manually vectorized to create a reference. As reference shoreline, both an orthorectified Sentinel-2, acquired on 1 January 2017, and the Sentinel-1 on 31 December 2016 itself were used. Due to the inherent characteristics of the SAR images, a better reference line can derive from orthoimagery. This is the reason why the Sentinel-2 optical image was also used as reference. As it was given already orthorectified, it was pre-processed by applying the atmosphere correction using the Sen2Cor tool in SNAP, and then, it was resampled to 10 m. Finally, for the same reason previously explained in Section 3.1, the NIR band (that is band 8 in Sentinel-2) was chosen to extract the reference shoreline. Then, two edge detection algorithms were tested. The first one is the Canny edge detector, which detects a wide range of edges in raster images and produces thin edges as a raster map. It operates, at first, with a Gaussian filter (based on normal distribution) to reduce the noise, then two orthogonal gradient images are computed. Finally, only relevant or significant edges are extracted by thresholding with hysteresis [44]. The result is shown in Figure 4.

Figure 4. Canny algorithms applied on the pre-processed Sentinel-1 image acquired on 31 December 2019.

Then, the J-Net Dynamic algorithm was tested. The result provided by the J-Net Dynamic algorithm on the SAR image is shown in Figure 5.

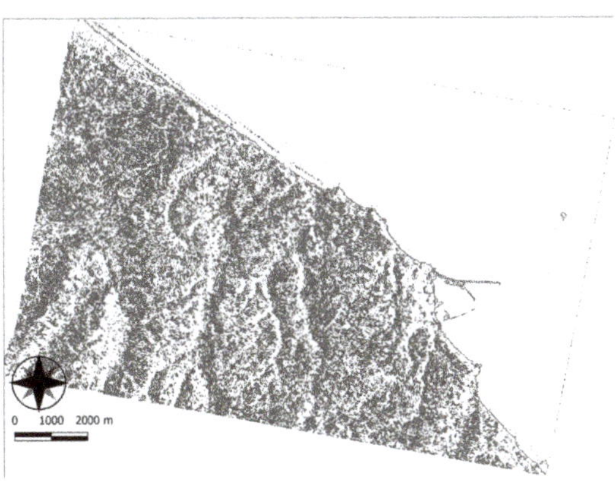

Figure 5. Application of the J-Net Dynamic algorithm on the SAR Sentinel-1 image acquired on 31 December 2016.

4. Results and Discussion

The shorelines extracted from the WorldView-2 image are presented in Figure 6: the red line illustrates the shoreline extracted manually, while the yellow one is the one generated using the WVWI and the blue one from the J-Net Dynamic.

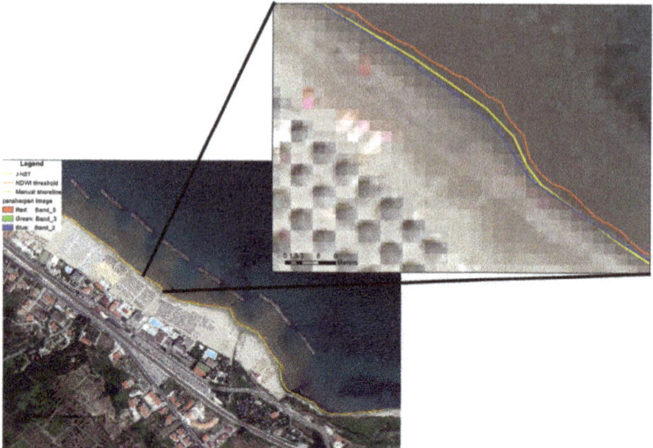

Figure 6. WorldView-2 image with three shorelines: the red line is the manually extracted shoreline; the yellow one is generated using the WVWI and the blue one from the J-Net Dynamic algorithm.

The three instantaneous shorelines extracted using the Sentinel-1 image are shown in Figure 7: the red one is the shoreline extracted manually, the yellow one is from the Canny algorithm and the blue one is from the J-Net Dynamic algorithm. The additional green line is the shoreline manually extracted from Sentinel-2, acquired on 1 January 2017; that is, one day after the Sentinel-1. It was also used as a reference for the reasons explained in Section 3.2.

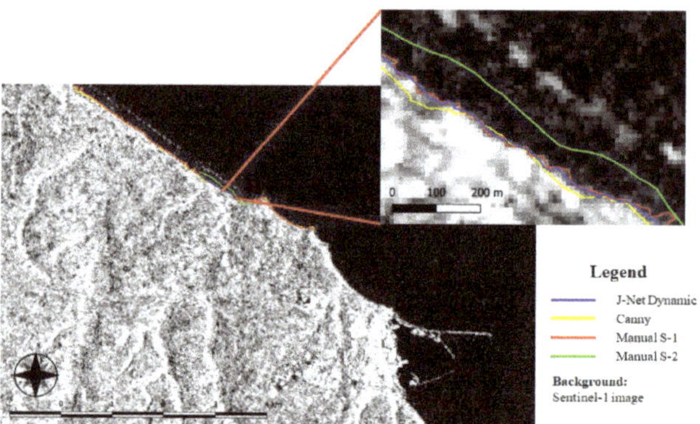

Figure 7. SAR Sentinel-1 image on the background showing four shorelines: the red and the green lines are, respectively, the shorelines manually extracted from Sentinel-1 and Sentinel-2 images; the yellow one is the one generated using the Canny algorithm and the blue one from the J-Net Dynamic algorithm.

In order to compare the accuracy of the obtained shorelines, an index I (2), reported in literature [16,45], is used:

$$I = \frac{S}{L},$$ (2)

where S is the area calculated between the shoreline extracted from the algorithm and the shoreline used as reference; and L is the length of the reference line. The index is expressed in meters. The values obtained are shown in Table 2.

Table 2. I Index values. Comparisons.

Image Type	Optical Image		SAR Image			
Reference shoreline extracted from	Pansharpened WV-2 image		Sentinel-1 image		Sentinel-2 image	
Algorithm	NDVI	J-Net Dynamic	Canny	J-Net Dynamic	Canny	J-Net Dynamic
I (m)	1.67	1.20	12.52	5.86	66.49	60.77
% increase		28.14		58.19		8.60

The results of the index indicate that the use of J-Net Dynamic algorithm improves the shoreline extraction both in the optical and in the SAR images compared to the common filters. For the optical WorldView-2 image, the index I is equal to 1.67 m using WVWI indexed image and 1.20 m using the J-Net Dynamic derived image. This means that, in the VHR image, the J-Net Dynamic algorithm improves the accuracy by about 28%. For the SAR Sentinel-1 image, considering as a reference the shoreline extracted from the Sentinel-1 image, the index I is equal to 12.52 m using Canny derived image and 5.86 m with the application of the J-Net Dynamic filter. In brief, in the HR SAR image, the accuracy is increased by about 58%. Considering as reference the shoreline extracted from Sentinel-2 image, the index I is 66.49 m for the Canny and 60.77 m for the J-Net Dynamic, which means an increase of almost 9%. All the results proves that the shoreline extracted from the J-Net Dynamic derived image is over-fitted to the reference compared to the ones extracted from the common filters. Two more things can be also noted: the first one is that there is a higher difference between the SAR image values and the optical image values. This happens because the SAR image has a lower resolution. The second one is that, with respect to the SAR image, there is a big difference depending on the chosen reference shoreline. The ones extracted with the algorithms are further away from the Sentinel-2 than the Sentinel-1 reference shoreline. The SAR image, due to its nature, could have little backscattering from the wet sand, which means that, in this case, it "confuses" the wet sand with water. However, the purpose of this paper is to test the J-Net Dynamic on optical and SAR images and to compare the results to the common algorithms. In all of the aforementioned cases, it proved to extract a shoreline closest to the reference one.

To better determine the accuracy of the shoreline extraction, a method already used in literature [10,46] was used. Ten transects (Figure 8), about one every 100 m and drawn from the land to the sea, were used to calculate the intersection between them and the seven shorelines extracted from WorldView-2, WVWI, J-Net Dynamic on the VHR image, Sentinel-1, Sentinel-2, Canny and J-Net Dynamic on the HR image.

Figure 8. Ten transects every 100 m, drawn from the land to the sea, to calculate the intersection between them and seven shorelines extracted from WorldView-2, WVWI, J-Net Dynamic on the VHR image, Sentinel-1, Sentinel-2, Canny and J-Net Dynamic on the HR image.

Then, the differences between the reference shorelines and the ones extracted from the algorithms were measured. In the end, the mean values and the standard deviations of every difference were calculated (Table 3).

Table 3. Statistic table. The means and the standard deviations have been calculated on the difference between the reference shorelines with the algorithms we tested. The difference has been calculated on the values obtained from the intersection between the transects and the seven shorelines.

	VHR OPTICAL			HR SAR		
	WV2 - WVWI	*WV2 - J Net*	*S2 - Canny*	*S2 - J Net*	*S1 - Canny*	*S1 - JNet*
Mean (m)	1.87	1.45	70.11	66.09	17.21	10.61
Standard deviation (m)	1.13	1.12	29.20	19.57	9.10	8.21

As shown in Table 3, for the VHR image, the standard deviation is 1.13 m and the mean is 1.87 considering the common filter. On the other hand, the standard deviation of the J-Net Dynamic is 1.12 m and the mean value 1.45 m. For the HR image, considering the Sentinel-2 optical image as a reference, the results show that the mean value of the difference with the Canny derived shoreline is 70.11 m and the standard deviation is 29.20 m. Compared with the J-Net values, that are 66.09 m and 19.57 m respectively, they are higher. The same happens considering the Sentinel-1 as a reference: the mean values for Canny and J-Net are 17.21 m and 10.61 m and the standard deviations are 9.10 m and 8.21 m, respectively. It can be noticed that the mean and standard deviation are lower for the VHR optical image than the HR SAR image, as expected. Moreover, the values obtained for the SAR image using two different reference shorelines confirm the better results obtained by using the Sentinel-1 image as references, as they were found previously. Indeed, the standard deviation for the difference between Sentinel-2 and Canny shorelines is 29.20 m, compared to 9.10 m obtained by the difference between the Sentinel-1 and Canny shorelines. The same happens for the differences between the Sentinel-2 with the J-Net (19.57 m) and Sentinel-1 with the J-Net (8.21 m). However, in every case, the J-Net Dynamic gives best accuracy compared to the common filters, as already found out with the index I.

In [16], the index I was used to compare the results obtained for the shoreline extraction using the NDVI and the NDWI (normalized difference water index, which considers the same bands of WVWI and the same equation) filters on the WorldView-2 initial and pansharpened image. It confirms that NDWI application provides better results; and that the pan-sharpening permits one to enhance geometric resolution and to reduce, to less than 1 m, the mean value of shifts between the automatically extracted shoreline and the manually vectorized one. They found an index value of 1.386 m for the multispectral image and 0.955 m for the pan-sharpened one. The better accuracy, compared to the one obtained for the research of this paper, that is, 1.67 m, could be due to the fact that they identified a different threshold to detect the shoreline based on a classification method using three contrasting classes (sea, land and vegetation) or due to the fact that the range of shoreline was both sandy and rocky. The rocky parts are less sensitive to instantaneous dynamic change than the sandy ones. The index I was also considered to evaluate the average shift between two temporal consecutive shorelines [11]. They compared shorelines during 2005 and 2011 and they evidenced that the effects of the human intervention, like the installation of breakwaters, limit erosion phenomena near shore zone. This tendency was also remarked on for the years 2011–2012.

The index I has not been used yet in literature to assess the SAR images. In [47], an automated method of the shoreline position detection using Sentinel-1 SAR image was studied. The method aimed at the automatic extraction of shoreline edge from pre-processed images. The algorithm performs four steps: despeckling, binarisation, morphological operations and edge detection by means of the Canny edge detector. Once the shoreline is extracted, it is compared to the one extracted from the images acquired by video system. The VDS (video-derived shoreline) is a collection of shoreline points. So, the distance between each point from the SDS (satellite-derived shoreline) is calculated in order to evaluate quantitatively the correctness of the SDS position. The average offset turns out to be about 10 m and the root mean square difference (RMSD) is 12.48 m.

To compare the accuracy with the case study discussed in this paper, the vertices of the reference shoreline (so the shoreline manually outlined on the pre-processed Sentinel-1 image) are extracted and the shortest distance between them and the Canny derived shoreline is calculated. The results show that the average offset is around 16 m, while the standard deviation is 11.40 m. So, a robust contrast can be performed also by applying the binarisation, as proposed by [47], because it splits exactly water and land, and as consequence, the edge detector can easily extract the contour of the image. To increase the accuracy, [48] propose an integration between RASAT pansharpened image and Sentinel-1 SAR image to improve the quality of the results. The first land/water segmentation was obtained using RASAT image by means of random forest classification method. Then, the result was used as training data set to define fuzzy parameters for shoreline extraction from Sentinel-1A SAR image. The manually digitized shoreline was used as a reference and the accuracy assessment was performed by calculating perpendicular distances between reference data and extracted shoreline by the aforementioned method. The mean distance value between the final result and the reference data was calculated to be 5.59 m, which is half pixel size of Sentinel-1A and which is almost three times better then the results obtained in this research. The higher accuracy was due not only to the integration between optical and SAR images, but also to the fact that the coast considered in [48] is mainly stony rather than sandy.

Due to the high resolution difference, the authors find it inappropriate to compare the WorldView-2 optical image and the Sentinel-1 SAR image. The purpose of the paper was only to discuss potentiality of the J-Net Dynamic algorithm both in optical and in SAR images independently. So, an additional test was performed. To compare the accuracy with the average offset obtained by means of the Canny edge detector (16 m with an RMSD equal to 11.40), the vertices of the pre-processed Sentinel-1 image shoreline were extracted and the shortest distance between them and the J-Net derived shoreline was calculated. An average offset around 7 m and a standard deviation of 7.36 m were found. The accuracy was lower than one pixel. The values confirm the trend that was delineated by means of the index I value: the algorithm J-Net Dynamic increased the accuracy of the extraction.

In future studies, other images, comparable in terms of resolution, such as Sentinel-1 and Sentinel-2 or WorldView-2 and COSMO-SkyMed images, can be integrated to improve the accuracy of the results. Moreover, to answer to the question of whether the Sentinel-1 is capable of monitoring coastal changes, another interesting future work could consider all the Sentinel-1 image data set to study the shoreline changes during a determined period of time.

5. Conclusions

Monitoring coastal areas with shoreline mapping allows one to analyze erosion and deposition phenomena in order to be able to propose corrective actions. In this work, an experimental algorithm, called J-Net Dynamic, has been presented to automatically extract the shoreline. It has been tested both in multispectral and in SAR images with different resolutions. Both images were pre-processed to improve their geometric and radiometric properties in order to facilitate their subsequent analysis. They were then processed with filters and indices commonly used for this kind of research, like the WorldView water index for the optical image and the Canny edge detector for the radar image, and the J-Net Dynamic. The J-Net Dynamic algorithm was mainly developed and previously applied in medical fields and it has been tested on a VHR optical image, and for the first time, also on an HR SAR image for environmental monitoring. To validate and verify the accuracy and robustness of this approach, a comparison with other typical algorithms was performed through an index I, which divides the area between the shoreline obtained by the algorithm and the reference one with the length of the latter, and through the use of ten transects along the shorelines. The results showed that J-Net Dynamic improves the detection, obtaining a shoreline closest to the reference, both for VHR and HR satellite images.

Author Contributions: These authors contributed equally to this work. All authors have read and agreed to the published version of the manuscript.

Funding: This research received no external funding.

Conflicts of Interest: The authors declare no conflict of interest.

Abbreviations

The following abbreviations are used in this manuscript:

VHR	very high resolution
HR	high resolution
SAR	synthetic aperture radar
GNSS	Global Navigation Satellite System
UAV	unmanned aerial vehicle
WI	water index
VI	vegetation index
ACM	active connections matrix
UTM	universal transverse mercator
WGS84	World Geodetic System 1984
EPSG	European Petroleum Survey Group
EC	European Commission
ESA	European Space Agency
SLC	single look complex
GRD	ground range detected
IW	Interferometric Wide
TOPSAR	Terrain Observation with Progressive Scans SAR
MLLW	mean Lower low water
RPC	rational polynomial coefficient
GCP	ground control point
CP	check point
HCS	hyperspherical color space
WVWI	WorldView water index
CB	coastal band
NIR	near infrared
SNAP	sentinel application platform
DN	digital number
NDVI	normalized difference vegetation index
NDWI	normalized difference Water Index
VDS	video-derived shoreline
SDS	satellite-derived shoreline
RMSD	root mean square difference
COSMO-SkyMed	COnstellation of small Satellites for the Mediterranean basin Observation

Appendix A

J-Net Dynamic

This algorithm was developed by Professor Paolo Massimo Buscema, mathematician, Director of the Semeion Research Centre of Science of Communication of Rome and Full Professor Adjoint at the University of Colorado (USA).

The patent concerns active connections matrix systems (ACM), according to which each image is considered as an active matrix (network) of connected elements (pixels) that develops over time. The main idea upon which this theory is based states that each digital image stores the maximum amount of information within the pixel values and their relationships. Furthermore, it is possible to obtain important information by analyzing the reciprocal positions occupied by pixels. For a complete presentation of ACM algorithms, see [18,49].

Any digital image is a matrix made of as many rows as the pixels that determine the width and as many columns as the pixel number related to the height. Each pixel is identified by its coordinates $i \in 1, \dots, R$, $j \in 1, \dots, C$, and its brightness $L \in 2^M$ (e.g., in the case of 256 shades of gray, M = 8). For each pixel u_{ij} a set containing all the surrounding pixels $I_{(u_{ij})}$ named *neighborhood* can be defined. In the ACM systems, all the pixels $u_{xy} \in I_{(u_{ij})}$ are linked to the central pixel u_{ij} by means of the $w_{(ij),(xy)}$ connections. The systems are classified into three orders of complexity, according to the type of evolution over time. In the first order of complexity, the values of connections are initialized once, at the beginning, and then remain fixed while pixel values $u_{ij}^{[n]}$ evolve over time until convergence, starting from the $u_{ij}^{[0]}$ value, directly derived from the image. The situation is specular in the case of second order, where the pixels' values are fixed and equal to $u_{ij}^{[0]}$, while the connections values $w_{(ij),(xy)}^{[n]}$ are updated at each iteration, after initializing them to values $w_{(ij),(xy)}^{[0]} \approx 0$. Finally, the third order of complexity includes models in which both the pixels and the connections change over time.

J–Net Dynamic [19,49] is a ACM system with dynamic connections and units (third order of complexity). The main prerogative of this method is to consider in its equations not only the central pixel as such, with its relative neighborhood, but also as part of the surroundings of each of the pixels around it, when they are considered in turn as the central pixel. At the beginning of the process, the units u_{ij} are linearly scaled into the range $[-1 + \alpha, 1 + \alpha]$, where α is a parameter to be set up. By varying the α value it is possible to study images in an iterative manner. The first part of computations involves exclusively the central pixel and its neighborhood I_{ij}, as shown by Equations (A2)–(A7). J–Net follows the schema: update of weights (Equations (A1)–(A5), computation of the new pixel values based on the weights (Equation (A6)), update of units (Equations (A7)–(A14), re-update of weights and so on. In this case, the update of units also involves the neighborhoods I_{xy} such that $u_{xy} \in I_{ij}$.

$$S_{ij}^{[n]} = \pi \cdot (r_{ij}^{[n]})^2 \tag{A1}$$

$$D_{ij}^{[n]} = \sum_{(xy)s.t.u_{xy} \in I_{ij}} (u_{ij}^{[n]} - w_{(ij),(xy)}^{[n]}) \tag{A2}$$

$$J_{ij}^{[n]} = \frac{e^{D_{ij}^{[n]}} - e^{-D_{ij}^{[n]}}}{e^{D_{ij}^{[n]}} + e^{-D_{ij}^{[n]}}} \tag{A3}$$

$$\Delta w_{(ij),(xy)}^{[n]} = -(u_{ij}^{[n]} - J_{ij}^{[n]}) \cdot (-2 \cdot J_{ij}^{[n]}) \cdot (1 - (J_{ij}^{[n]})^2) \cdot (u_{xy}^{[n]} - w_{(ij),(xy)}^{[n]}) \tag{A4}$$

$$w_{(ij),(xy)}^{[n+1]} = w_{(ij),(xy)}^{[n]} + \Delta w_{(ij),(xy)}^{[n]} \tag{A5}$$

$$P_{ij}^{[n]} = S_P \cdot \frac{1}{|I_{ij}|} \cdot \left(\sum_{(xy)s.t.u_{xy} \in I_{ij}} w_{(ij),(xy)}^{[n]} \right) + O_P, \tag{A6}$$

where:

$$S_P = \frac{M_P}{M_w - m_w}$$

$$O_P = -\frac{m_w \cdot M_P}{M_w - m_w}.$$

M_P denotes the maximum value available for pixels, M_w and m_w are, respectively, the maximum and the minimum weights $w_{(ij),(xy)}^{[n]}$ $\forall i, x \in \{1, \dots, R\}$ and $j, y \in \{1, \dots, C\}$.

$$Out_{ij}^{[n]} = S_0 \cdot \frac{1}{|I_{ij}|} \cdot \left(\sum_{(xy)s.t.u_{xy} \in I_{ij}} w_{(ij),(xy)}^{[n]} \right) + O_0 \tag{A7}$$

where, with the usual notation:

$$S_0 = \frac{2}{M_w - m_w}$$

$$O_0 = -\frac{M_w + m_w}{M_w - m_w}.$$

The internal activation state of each pixel is then defined as:

$$S_{ij}^{[n]} = |Out_{ij}^{[n]}|. \tag{A8}$$

Therefore, the closer the weighted average of the connections of each pixel with those of its surroundings is to a neutral value, 0 or 127 depending on the encoding, the higher the value of the internal activation state of the pixel itself. With the aim of defining an update rule for units, the quantity $\Delta S_{(ij),(xy)}^{[n]}$ is considered (Equation (A9)). Then, according to Equations (A10)–(A11), the transition from I_{ij} to I_{xy} takes place.

$$\Delta S_{(ij),(xy)}^{[n]} = -tanh(S_{ij}^{[n]} - u_{xy}^{[n]}) \tag{A9}$$

$$\varphi_{ij}^{[n]} = L \cdot u_{ij}^{[n]} \cdot \sum_{(xy)s.t.u_{xy} \in I_{ij}} \left(1 - \left(\Delta S_{(xy),(ij)}^{[n]}\right)^2\right) \tag{A10}$$

$$\psi_{ij}^{[n]} = \sum_{(xy)s.t.u_{xy} \in I_{ij}} tanh(\varphi_{xy}^{[n]}). \tag{A11}$$

The delta quantities required for correction shall be calculated in the last step. It is possible to choose for two different update laws named union (Equation (A12)) and intersection (Equation (A13)).

$$\delta u_{ij}^{[n]} = \varphi_{ij}^{[n]} + \psi_{ij}^{[n]} \tag{A12}$$

$$\delta u_{ij}^{[n]} = \varphi_{ij}^{[n]} \cdot \psi_{ij}^{[n]} \tag{A13}$$

$$u_{ij}^{[n+1]} = u_{ij}^{[n]} + \delta u_{ij}^{[n]}. \tag{A14}$$

References and Note

1. Łabuz, T.A. Environmental impacts-coastal erosion and coastline changes. In *Second Assessment of Climate Change for the Baltic Sea Basin*; Springer: Cham, Switzerland, 2015; pp. 381–396.
2. Carapuço, M.M.; Taborda, R.; Silveira, T.M.; Psuty, N.P.; Andrade, C.; Freitas, M.C. Coastal geoindicators: Towards the establishment of a common framework for sandy coastal environments. *Earth-Sci. Rev.* **2016**, *154*, 183–190. [CrossRef]
3. Dolan, R.; Hayden, B.P.; May, P.; May, S. The reliability of shoreline change measurements from aerial photographs. *Shore Beach* **1980**, *48*, 22–29.
4. Sorensen, R.M. *Basic Coastal Engineering*; Springer Science & Business Media: New York, NY, USA, 2005; Volume 10.
5. Gens, R. Remote sensing of coastlines: Detection, extraction and monitoring. *Int. J. Remote Sens.* **2010**, *31*, 1819–1836. [CrossRef]
6. Toure, S.; Diop, O.; Kpalma, K.; Maiga, A.S. Shoreline Detection using Optical Remote Sensing: A Review. *ISPRS Int. J. Geo-Inf.* **2019**, *8*, 75. [CrossRef]
7. Boak, E.H.; Turner, I.L. Shoreline definition and detection: A review. *J. Coastal Res.* **2005**, 688–703. [CrossRef]
8. Pugliano, G.; Robustelli, U.; Luccio, D.D.; Mucerino, L.; Benassai, G.; Montella, R. Statistical Deviations in Shoreline Detection Obtained with Direct and Remote Observations. *J. Mar. Sci. Eng.* **2019**, *7*, 137. [CrossRef]
9. Palazzo, F.; Latini, D.; Baiocchi, V.; Del Frate, F.; Giannone, F.; Dominici, D.; Remondiere, S. An application of COSMO-Sky Med to coastal erosion studies. *Eur. J. Remote Sens.* **2012**, *45*, 361–370. [CrossRef]

10. Dominici, D.; Zollini, S.; Alicandro, M.; Della Torre, F.; Buscema, P.M.; Baiocchi, V. High Resolution Satellite Images for Instantaneous Shoreline Extraction Using New Enhancement Algorithms. *Geosciences* **2019**, *9*, 123. [CrossRef]

11. Maglione, P.; Parente, C.; Vallario, A. High resolution satellite images to reconstruct recent evolution of domitian coastline. *Am. J. Appl. Sci.* **2015**, *12*, 506. [CrossRef]

12. Liu, H.; Jezek, K.C. Automated extraction of coastline from satellite imagery by integrating Canny edge detection and locally adaptive thresholding methods. *Int. J. Remote Sens.* **2004**, *25*, 937–958. [CrossRef]

13. Zhang, T.; Yang, X.; Hu, S.; Su, F. Extraction of coastline in aquaculture coast from multispectral remote sensing images: Object-based region growing integrating edge detection. *Remote Sens.* **2013**, *5*, 4470–4487. [CrossRef]

14. Sheng, G.; Yang, W.; Deng, X.; He, C.; Cao, Y.; Sun, H. Coastline detection in synthetic aperture radar (SAR) images by integrating watershed transformation and controllable gradient vector flow (GVF) snake model. *IEEE J. Ocean. Eng.* **2012**, *37*, 375–383. [CrossRef]

15. Ma, H.; Zhang, L. Ocean SAR Image Segmentation and Edge Gradient Feature Extraction. *J. Coast. Res.* **2019**, *94* (Suppl. S1), 141–144. [CrossRef]

16. Maglione, P.; Parente, C.; Vallario, A. Coastline extraction using high resolution WorldView-2 satellite imagery. *Eur. J. Remote Sens.* **2014**, *47*, 685–699. [CrossRef]

17. McFeeters, S.K. The use of the Normalized Difference Water Index (NDWI) in the delineation of open water features. *Int. J. Remote Sens.* **1996**, *17*, 1425–1432. [CrossRef]

18. Buscema, P.M. *Sistemi ACM e Imaging Diagnostico: Le Immagini Mediche Come Matrici Attive di Connessioni*; Springer Science & Business Media: New York, NY, USA, 2006.

19. Buscema, M.; Catzola, L.; Grossi, E. Images as active connection matrixes: The J-net system. *Int. J. Intell. Comput. Med. Sci. Image Process.* **2008**, *2*, 27–53. [CrossRef]

20. Le spiagge dell'Abruzzo. *Studi costieri.* 2006; Volume 10. Available online: http://www.gnrac.it/rivista/Numero10.htm (accessed on 2 December 2019).

21. Progetto Preliminare Della Costa Teatina—Fenomeni Erosivi Della Fascia Costiera. Available online: http://www.provincia.chieti.it/flex/cm/pages/ServeAttachment.php/L/IT/D/2%252F4%252F4%252FD.d367c153fd7837fd72ca/P/BLOB%3AID%3D3926/E/pdf (accessed on 28 October 2019).

22. MATTM-Regioni. Linee Guida per la Difesa della Costa dai fenomeni di Erosione e dagli effetti dei Cambiamenti climatici. Versione 2018. Available online: http://www.erosionecostiera.isprambiente.it/linee-guida-nazionali (accessed on 29 October 2019).

23. European Space Imaging. Available online: https://www.euspaceimaging.com/about/satellites/worldview-2/ (accessed on 19 September 2019).

24. Satellite Imaging Corporation. Available online: https://www.satimagingcorp.com/satellite-sensors/worldview-2/ (accessed on 19 September 2019).

25. Geoimage. Available online: https://www.geoimage.com.au/satellite/worldview-2 (accessed on 19 September 2019).

26. Geomatics. Available online: https://www.https://geomatics.planet.com/upload/digitalglobe/DigitalGlobe%20Core%20Imagery%20Products%20Guide.pdf (accessed on 19 September 2019).

27. ESA Sentinel Home, Missions, Sentinel-1. Available online: https://sentinel.esa.int/web/sentinel/missions/sentinel-1/overview (accessed on 22 May 2019).

28. Copernicus Open Access Hub. Available online: https://scihub.copernicus.eu/dhus/#/home (accessed on 22 May 2019).

29. ESA Sentinel Home, User Guides, Sentinel-1 SAR, Product Types and Processing Levels, Level-1. Available online: https://sentinel.esa.int/web/sentinel/user-guides/sentinel-1-sar/product-types-processing-levels/level-1 (accessed on 22 May 2019).

30. ESA Sentinel Home, User Guides, Sentinel-1 SAR, Acquisition Modes, Interferometric Wide Swath. Available online: https://sentinel.esa.int/web/sentinel/user-guides/sentinel-1-sar/acquisition-modes/interferometric-wide-swath (accessed on 24 May 2019).

31. TIDES4FISHING. Available online: https://tides4fishing.com/ (accessed on 29 March 2019).

32. Whitfield, A.; Elliott, M. 1.07-Ecosystem and biotic classifications of estuaries and coasts. In *Treatise on Estuarine and Coastal Science*; Elsevier: Amsterdam, The Netherlands, 2011; pp. 99–124.

33. Planetek, ERDAS IMAGINE, Software, PRODUCER Suite of Power Portfolio by Hexagon Geospatial.

34. Padwick, C.; Deskevich, M.; Pacifici, F.; Smallwood, S. WorldView-2 Pansharpening. In Proceedings of the ASPRS 2010 Annual Conference, San Diego, CA, USA, 26–30 April 2010.

35. Wolf, A.F. Using WorldView-2 Vis-NIR multispectral imagery to support land mapping and feature extraction using normalized difference index ratios. In *International Society for Optics and Photonics Algorithms and Technologies for Multispectral, Hyperspectral, and Ultraspectral Imagery XVIII*; International Society for Optics and Photonics: San Diego, CA, USA, 2012; Volume 8390.

36. Palomar-Vázquez, J.; Almonacid-Caballer, J.; Pardo-Pascual, J.E.; Sanchez-García, E. SHOREX: A new tool for automatic and massive extraction of shorelines from Landsat and Sentinel 2 imagery. In Proceedings of the 7th International Conference on the Application of Physical Modelling in Coastal and Port Engineering and Science (Coastlab), Santander, Spain, 22–26 May 2018.

37. Bodansky, E.; Gribov, A.; Pilouk, M. Smoothing and compression of lines obtained by raster-to-vector conversion. In *International Workshop on Graphics Recognition*; Springer, Berlin/Heidelberg, Germany, 2001; pp. 256–265.

38. Alesheikh, A.A.; Ghorbanali, A.; Nouri, N. Coastline change detection using remote sensing. *Int. J. Environ. Sci. Technol.* **2007**, *4*, 61–66. [CrossRef]

39. Braga, F.; Tosi, L.; Prati, C.; Alberotanza, L. Shoreline detection: Capability of COSMO-SkyMed and high-resolution multispectral images. *Eur. J. Remote Sens.* **2013**, *46*, 837–853. [CrossRef]

40. ESA, Earth Online, Home, Missions, ESA EO Missions, ERS, Instruments, SAR, Applications. Available online: https://earth.esa.int/web/guest/missions/esa-operational-eo-missions/ers/instruments/sar/applications/radar-courses/content-3/-/asset_publisher/mQ9R7ZVkKg5P/content/radar-course-3-image-interpretation-tone (accessed on 3 September 2019).

41. Marghany, M.; Sabu, Z.; Hashim, M. Mapping coastal geomorphology changes using synthetic aperture radar data. *Int. J. Phys. Sci.* **2010**, *5*, 1890–1896.

42. Pradhan, B.; Rizeei, H.; Abdulle, A. Quantitative assessment for detection and monitoring of coastline dynamics with temporal RADARSAT images. *Remote Sens.* **2018**, *10*, 1705. [CrossRef]

43. Taha, L.G.E.D.; Elbeih, S.F. Investigation of fusion of SAR and Landsat data for shoreline super resolution mapping: The northeastern Mediterranean Sea coast in Egypt. *Appl. Geom.* **2010**, *2*, 177–186. [CrossRef]

44. GRASS Manual. Available online: https://grass.osgeo.org/grass76/manuals/addons/i.edge.html (accessed on 5 July 2019).

45. Guastaferro, F.; Maglione, P.; Parente, C.; Santamaria, R. Estrazione in automatico della linea di costa da immagini satellitari IKONOS. In Proceedings of the Conference "Geomatica. Le Radici del Futuro. Tributo a Sergio Degual & Riccardo Galetto", Pavia, Italy, 10–11 February 2011; pp. 109–116, ISBN 88-901939-6-4.

46. Esmail, M.; Mahmod, W.E.; Fath, H. Assessment and prediction of shoreline change using multi-temporal satellite images and statistics: Case study of Damietta coast, Egypt. *Appl. Ocean Res.* **2019**, *82*, 274–282. [CrossRef]

47. Spinosa, A.; Ziemba, A.; Saponieri, A.; Navarro-Sanchez, V.D.; Damiani, L.; El Serafy, G. Automatic Extraction of Shoreline from Satellite Images: A new approach. In Proceedings of the 2018 IEEE International Workshop on Metrology for the Sea; Learning to Measure Sea Health Parameters (MetroSea), Bari, Italy, 8–10 October 2018; pp. 33–38.

48. Demir, N.; Oy, S.; Erdem, F.; Şeker, D.Z.; Bayram, B. Integrated shoreline extraction approach with use of Rasat MS and SENTINEL-1A SAR Images. *ISPRS Ann. Photogramm. Remote Sens. Spat. Inf. Sci.* **2017**, *4*, 445. [CrossRef]

49. Buscema, M.; Grossi, E. J-Net System: A New Paradigm for Artificial Neural Networks Applied to Diagnostic Imaging. In *Applications of Mathematics in Models, Artificial Neural Networks and Arts*; Publishing House: Dordrecht, The Netherlands, 2010; pp. 431–455.

Article

Automatic Shoreline Detection from Eight-Band VHR Satellite Imagery

Maria Alicandro [1], Valerio Baiocchi [2], Raffaella Brigante [3] and Fabio Radicioni [3,*]

[1] Department of Civil, Construction-Architecture and Environmental Engineering, University of dell'Aquila, 67100 L'Aquila, Italy; maria.alicandro@univaq.it
[2] Department of Civil Construction and Environmental Engineering, Sapienza University of Rome, 00185 Rome, Italy; valerio.baiocchi@uniroma1.it
[3] Department of Engineering, University of Perugia, 06125 Perugia, Italy; raffaella.brigante@yahoo.it
* Correspondence: Fabio.radicioni@unipg.it

Received: 12 November 2019; Accepted: 11 December 2019; Published: 13 December 2019

Abstract: Coastal erosion, which is naturally present in many areas of the world, can be significantly increased by factors such as the reduced transport of sediments as a result of hydraulic works carried out to minimize flooding. Erosion has a significant impact on both marine ecosystems and human activities; for this reason, several international projects have been developed to study monitoring techniques and propose operational methodologies. The increasing number of available high-resolution satellite platforms (i.e., Copernicus Sentinel) and algorithms to treat them allows the study of original approaches for the monitoring of the land in general and for the study of the coastline in particular. The present project aims to define a methodology for identifying the instantaneous shoreline, through images acquired from the WorldView 2 satellite, on eight spectral bands, with a geometric resolution of 0.5 m for the panchromatic image and 1.8 m for the multispectral one. A pixel-based classification methodology is used to identify the various types of land cover and to make combinations between the eight available bands. The experiments were carried out on a coastal area with contrasting morphologies. The eight bands in which the images are taken produce good results both in the classification process and in the combination of the bands, through the algorithms of normalized difference vegetation index (NDVI), normalized difference water index (NDWI), spectral angle mapper (SAM), and matched filtering (MF), with regard to the identification of the various soil coverings and, in particular, the separation line between dry and wet sand. In addition, the real applicability of an algorithm that extracts bathymetry in shallow water using the "coastal blue" band was tested. These data refer to the instantaneous shoreline and could be corrected in the future with morphological and tidal data of the coastal areas under study.

Keywords: WorldView-2; Abruzzo; multispectral classification; shoreline; coastline

1. Introduction

The coastal environment is an extraordinary natural and economic resource that is subject to continuous transformation. The coastal area is a highly dynamic system where erosion and deposition are influenced by various factors, including meteorological, biological, geological, and anthropogenic factors. The theoretical definition of coastline is merely the transition between the sea and the land [1–3]. In theory, the concept is very simple and intuitive, but its application is actually a complex task because of the temporal variability of the coastline itself. This variability develops on scales profoundly different, from instantaneous to secular variations, and it depends on various factors including wave motion, tides, winds, erosion, and deposition. It has been noted that the most significant and potentially incorrect assumption in many shoreline investigations is that the instantaneous shoreline represents "normal" or "average" conditions [3]. For this reason, what it is important to underline that what is obtained from survey in the field or remote sensing are generally indicators of the actual coastline [3–5]. According to some authors, there are three main subdivisions of coastline indicators: characteristics visible by an operator on an aerial or remote sensing image; intersection between a tidal datum and a digital terrain model or a coastal profile; and characteristics of multispectral images identified by automatic algorithms not necessarily visible to unaided operator [3]. In this work, we assess techniques associated with the third type by comparing them with each other and with those of the first type. The purpose of this work is to assess potentialities of automatic extraction of the coastline in a specific area particularly prone to erosion. In fact, the Abruzzo region (Italy) coastline has a length of 125 km, of which 26 km is high coasts and 75 km is sandy coasts; the latter, therefore, are 80% of the entire coastline, and more than 50% are under erosion effects. The study of this paper was conducted on the sandy coast of Ortona (Abruzzo Region), in which three erosion phenomena have been reported on the coast, four active landslides have effects on the coast, and three landslide crags occur on the sea [6]. For this reason, according to the guidelines for the defense of the coast against erosion and the effects of climate change [7], one of the fundamental elements of a coastal information system is the extraction of the shoreline. A specific assessment on these areas has not yet been carried out and its results will be very useful for the continuation of the research and for the applications that can be implemented. It is important to note that what is extracted are indicators of the coastline and not the coastline itself (Figure 1).

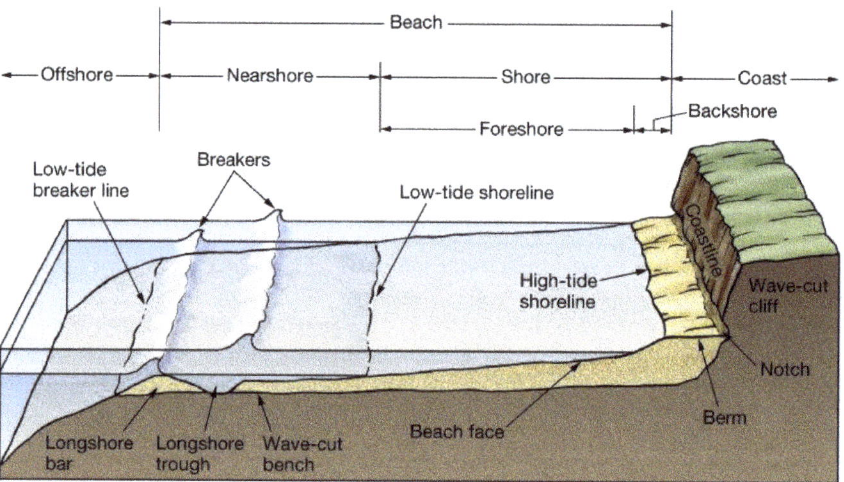

Figure 1. Indicators of the coastline.

In this work, after a brief illustration of the state of the art on the techniques of extraction of the coastline, we will introduce the materials and methods used for the experiments illustrated. Subsequently, the results obtained by them will be shown in the following specific paragraph, while the conclusions and possible future developments will be discussed together in the last paragraph of the paper.

2. State of the Art of Instant Shoreline Survey

From what has been said previously, it is easy to understand that the relief of the instantaneous shoreline is a complex feature due to the intrinsic dynamism of the shoreline itself. A wide variety of methodologies have been used [7–10] that can be summarized mainly in the following:

1. Photogrammetry/videography from airplane or UAV: In the first case, this is a matter of using the well-known techniques of photogrammetry from single acquisitions of aerial images that are currently almost all digital images; digital cameras can easily acquire films that can be treated with classical photogrammetric algorithms or with new approaches such as structure from motion (SFM) [11,12].
2. Terrestrial video systems: These are fixed camera systems which acquire at fixed time-intervals and which are composed of several cameras distributed along the coast with acquisition angles of up to 180 degrees. This technique produces oblique images that must be orthorectified [13].
3. Earth and satellite geomatic surveys: These are the classic surveys with instruments such as GPS/GNSS and total stations, levels that have a remarkable accuracy but can detect a limited number of points at different times.
4. Terrestrial and aerial lidar survey that can reconstruct both the surfaces of water and ground [12] or penetrate shallow water.
5. Remote sensing from satellite: It should be noted that with the development of remote sensing, shoreline detection is mainly achieved by image processing [14]. The availability of multispectral satellite images at very high resolution (VHR) allows, in fact, acquisition in a short time and simultaneously of long stretches of coast. The geometric accuracies of submetric to decimetric order are absolutely compatible with the specific application and the availability of different bands allows semi-automatic or automatic approaches [15–17] such as those proposed in this paper.

3. Materials and Methods

3.1. The WorldView-2 Satellite

The WorldView-2 satellite, launched in October 2009, added to the already existing constellation of commercial satellites DigitalGlobe [18], already formed at the time by the satellites WorldView-1 (launched in 2007) and QuickBird (launched in 2001). Its planned mission duration was 7.25 years, but it has just turned 10 years old. WorldView-2 operates at an altitude of 770 km with an inclination of 97.2° for a maximum orbital period of 100 min. It is equipped with instrumentation that allows it to collect high-resolution multispectral images and stereoscopic images.

The maximum ground resolution of WorldView-2 in the panchromatic band is 46 cm, while in the multispectral band it is 1.8 m, however the distribution and use of images with a resolution greater than 50 cm in the panchromatic band and 2 meters in the multispectral band is subject to approval by the US government. The high spatial resolution allows discrimination of details, such as vehicles, shallows, and individual trees in an orchard, while the high spectral resolution provides detailed information on different areas, such as road surface quality, sea depth, and plant health. The images are taken in eight spectral bands. In addition to the four standard bands (blue, green, red, and near infrared), WorldView-2 includes four new bands at 1.8 m resolution: coastal blue, yellow, red edge, and near infrared-2 (Figure 2).

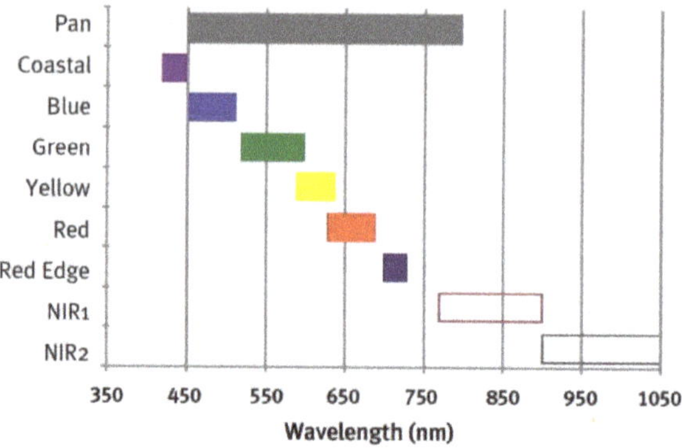

Figure 2. WorldView-2 spectral bands.

3.2. Test Site

The images used in this project are multispectral WorldView-2 images representing a part of the coast of the city of Ortona (Italy) (Figure 3). For this specific project, the tests were performed on a bundle of panchromatic and multispectral images acquired simultaneously in June. To be sure that the variability of the results depended only on the algorithm used in the various tests performed and not on factors due to different inclination of the image or the sun, the tests were all repeated on the same bundle of images. The images were processed entirely on the whole scene, but the results were highlighted on a single profile that was considered significant and representative of the tests on the whole scene. Ortona is a seaside town overlooking the Mediterranean Sea, more specifically, the Adriatic Sea (Figure 4); the coast is high and rocky in the southern part while it is low and sandy in the central northern part. The inner part of the city of Ortona is partly flat and partly hilly.

Figure 3. The test site area on the Adriatic coast (red circle).

Figure 4. Test site area: the yellow box is the area shown in the following figures.

In order to evaluate the possibilities of automatic extraction of the coastline (understood as an instantaneous shoreline as visible on satellite images), different combinations of algorithms on different spectra of the available images were tested and are illustrated below. The tests were developed in the ENVI 5.5 environment [19].

4. Experimentation and Results

4.1. Analysis of Individual Bands and Band Combinations

The first step in such an analysis is almost always pansharpening [20]; this operation creates a single multilayer image combining the panchromatic image of higher spatial resolution (in this case 50 cm) with the various multispectral levels (Figure 5) that have more thematic information but lower geometric resolution (in this case 1.8 m). Several algorithms have been developed to optimize pansharpening, and the comparison between them is at the center of a scientific debate [21–23].

In this experiment, Graham Smith's algorithm [24] was used because it has demonstrated to be the most reliable and accurate in previous studies [21,23]. This algorithm first calculates the weights of the single bands with the following equations:

$$B_{wt} = \int_{0.4}^{0.5} OT_B(\lambda) * SR_B(\lambda) * SR_P(\lambda) \tag{1}$$

$$G_{wt} = \int_{0.5}^{0.6} OT_G(\lambda) * SR_G(\lambda) * SR_P(\lambda) \tag{2}$$

$$R_{wt} = \int_{0.6}^{0.7} OT_R(\lambda) * SR_R(\lambda) * SR_P(\lambda) \tag{3}$$

$$NIR_{wt} = \int_{0.7}^{0.9} OT_{NIR}(\lambda) * SR_{NIR}(\lambda) * SR_P(\lambda) \tag{4}$$

where *OT* is optical transmittance and *SR* is the spectral response of the various bands with their wavelengths. Once the specific weights have been calculated, the simulated band is created using the following equation:

$$PanBand = (B * B_{wt}) + (G * G_{wt}) + (R * R_{wt}) + (NIR * NIR_{wt}) \tag{5}$$

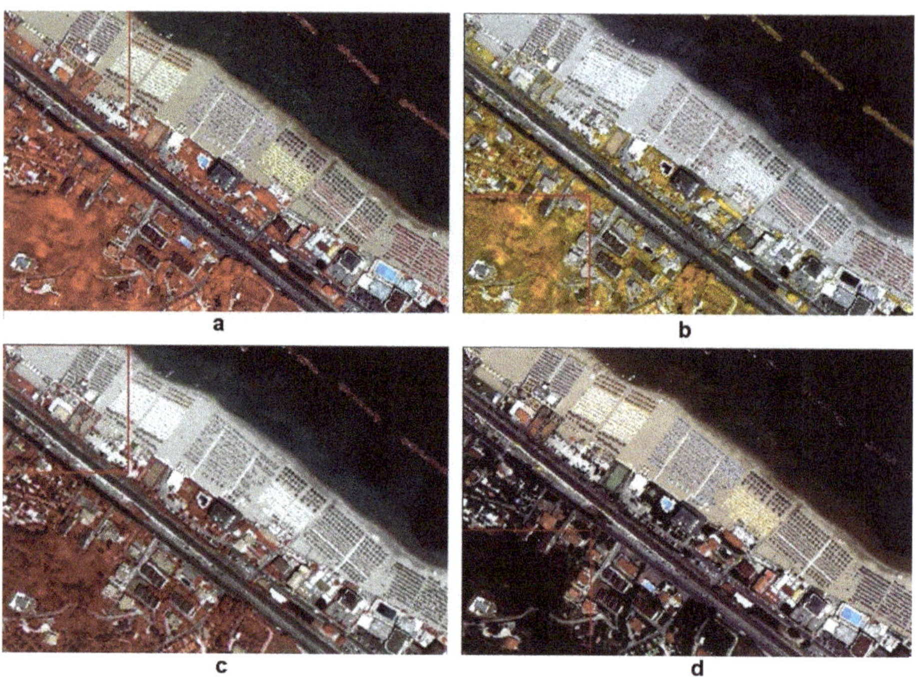

Figure 5. Some band combinations: (**a**) NIR1-G-B; (**b**) NIR1-Rededge-R; (**c**) R-G-B; (**d**) Rededge-R-Y.

Once the pansharpening has been undertaken (Figure 6), it is possible to start to perform operations on the single bands or on combinations of them.

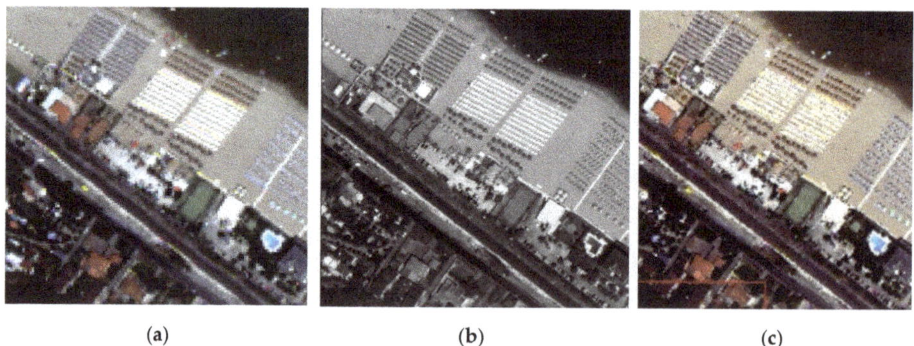

Figure 6. The pansharpening process: (**a**) MultiSpectral; (**b**) Panchromatic; (**c**) PanSharpened.

Among the most used indices in the combination of the bands there is undoubtedly the vegetation index (NDVI) and the normalized difference water index (NDWI). Both these indices have been tested because they automatically highlight the instantaneous shoreline, showing a remarkable discontinuity approaching it. The NDVI is a well-known index used to study vegetation, exploiting the different response between the near-infrared band and the red band:

$$NDVI = \frac{NIR - R}{NIR + R} \tag{6}$$

In vegetation studies, it is widely used because it allows to distinguish the plants that correctly perform the chlorophyll synthesis from those that do not. In vegetation applications it provides a dimensionless value between −1 and 1 but can also be usefully used for the coastline studies because the proximity of the shoreline assumes values between −0.5 and −0.1.

The second index, the NDWI, has been studied just to distinguish the areas covered by water from those that are not; it is actually very similar to the NDVI but uses the green band instead of the red one, and its formulation is in fact the following:

$$NDWI = \frac{NIR - G}{NIR + G} \tag{7}$$

Using the NDWI, the areas covered by water are characterized by positive values of the index while vegetation and bare soil usually show negative values. Dry sand, because of its high reflectance in the green band, usually shows positive values that are close to zero (Figure 7).

(a)   (b)

Figure 7. (**a**) Normalized difference vegetation index (NDVI) and (**b**) normalized difference water index (NDWI) processed images in gray scale.

4.2. Automatization of Water and Vegetation Detection

Below, we will analyze the results that can be obtained with the various algorithms available by comparing them on a specific profile that showed the characteristic aspects of a high coast and a sandy coast. On the panchromatic image, the sea/beach limits have been vectorized on the screen by an expert photo-interpreter. The purpose of this research is in fact to verify how efficient and accurate the automatic algorithms can be compared to the manual vectorization that is currently used to extract the instantaneous shoreline from satellite images. The ground surveys, already tested in previous researches [14], are certainly more accurate than those obtained from satellite images, but on sandy shores they cannot acquire the entire line of the instantaneous shoreline at the same time as it is visible on a satellite image and, therefore, are not directly comparable with them. The beach/sea and sea/detached-breakwater limits vectorized by the operator on the profile have been considered as reference for all the algorithms tested and are highlighted by the dashed lines in final comparison of the various algorithms tested. Considering that the pixel size in panchromatic is about 50 cm, the differences between the beach/sea limits vectorized by the operator with those extracted automatically equal to or less than 50 cm were not considered significant.

In this way, it will be possible to estimate the accuracy of the various algorithms, both in absolute and relative terms. What was important to observe was the possible discontinuity in the digital numbers of the images after the processing with the various algorithms; in fact, the edge-detector algorithms can easily trace border vector lines in the presence of sharp discontinuities.

The profile executed on the image processed with the NDVI algorithm shows a strong variation in slope at the boundaries of sea/beach and sea/detached-breakwater (Figure 8).

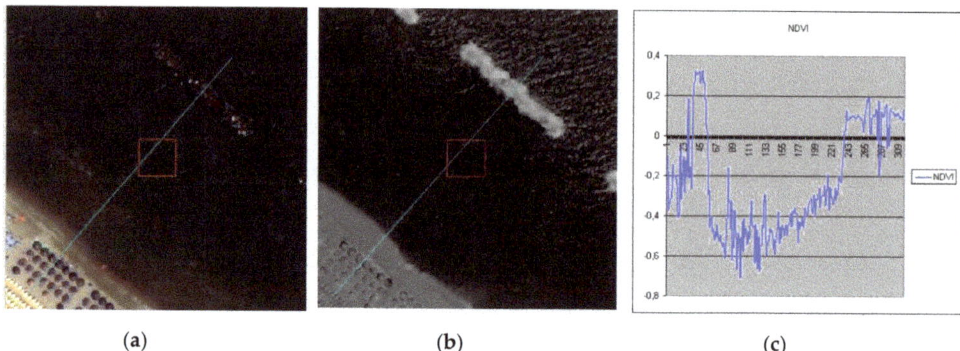

(a)   (b)   (c)

Figure 8. NE to SW NDVI profile path on (**a**) RGB and (**b**) NDVI processed images; (**c**) NE to SW NDVI profile, distances on the x-axis are expressed in metres. The profile trend shows sharp discontinuities at sea/land limits.

The spectral angle mapper (SAM) and matched filtering (MF) algorithms are very similar to NDVI but use all four bands (R, G, B, IR). SAM determines the spectral similarity between two spectra by calculating the angle between the spectra and treating them as vectors in a space with dimensionality equal to the number of bands. This technique, when used on calibrated reflectance data, is relatively insensitive to illumination and albedo effects [25]. MF, instead, maximizes the response of the known endmember and suppresses the response of the composite unknown background, thus matching the known signature [26] (Figure 9).

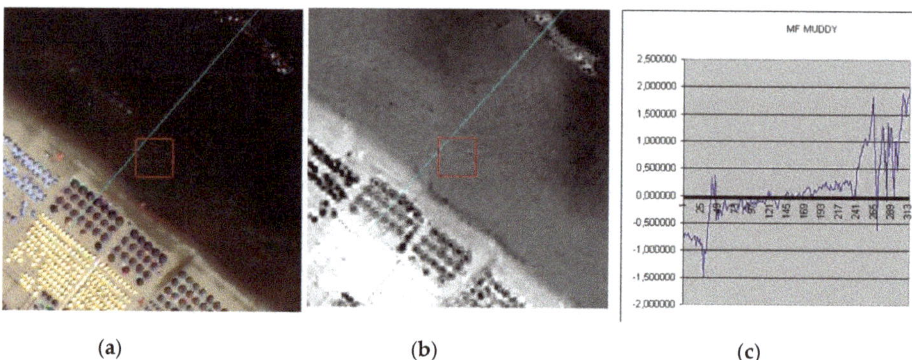

(a)   (b)   (c)

Figure 9. NE to SW matched filtering (MF) muddy profile path on (**a**) RGB and (**b**) MF muddy processed images; (**c**) NE to SW MF muddy profile, distances on the x-axis are expressed in metres. The profile trend does not show sharp discontinuities at sea/land limits.

The MF/SAM ratio has the characteristic of highlighting the transition from wet to dry sand that is revealed by a rapid change in the slope of the curve that represents it (Figure 10).

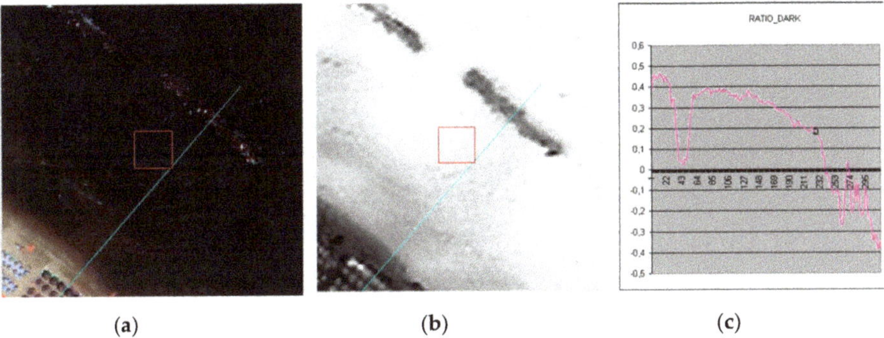

Figure 10. NE to SW MF/spectral angle mapper (SAM) profile path on (**a**) RGB and (**b**) MF/SAM processed images; (**c**) NE to SW MF/SAM profile, distances on the x-axis are expressed in metres. The profile trend shows visible discontinuities at sea/land limits.

The availability of the coastal blue band allowed us to exploit the algorithm called relative depth, developed by Stumpf and Holderied [27]. Using combinations of all bands provided a value of water depth up to about 14 m deep.

As the name of the algorithm says, the depth is relative, and in fact the values vary between 0 and 1; for bathymetric studies, therefore, control points measured to calibrate the model are required (Figures 11 and 12).

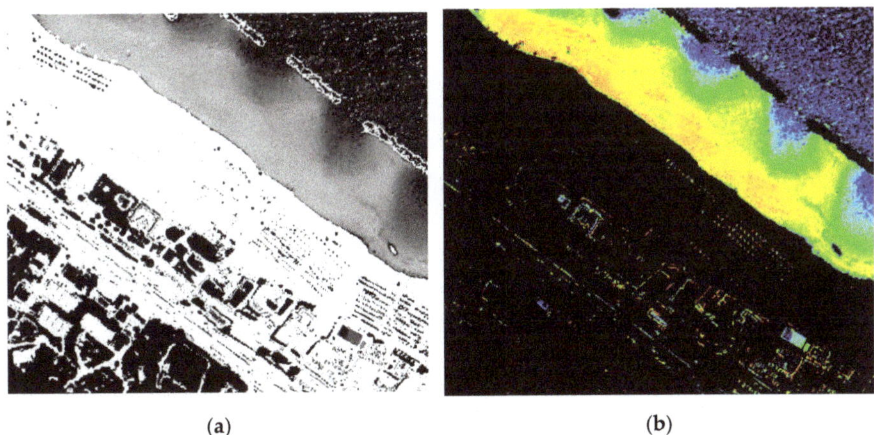

Figure 11. Relative depth algorithm results represented in (**a**) gray and (**b**) pseudocolor scales.

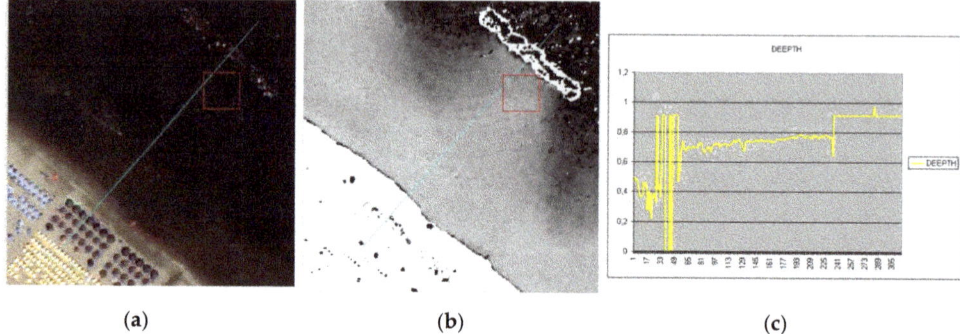

(a)   (b)   (c)

Figure 12. NE to SW relative depth profile path on (**a**) RGB and (**b**) relative depth processed images; (**c**) NE to SW relative depth profile, distances on the x-axis are expressed in metres. The profile trend shows almost vertical discontinuities at sea/land limits.

Comparing the various algorithms (Figure 13), it can be observed that they all show a marked discontinuity near the actual line of contact between sea and land; it is important to note that values in the y direction are reported for every specific algorithm so are not directly comparable, but what is important to observe is the discontinuity in the different trends. Among them, the most efficient and accurate algorithm seems to be the relative depth, because the discontinuity shown at the passage between water and land is the deepest, showing an almost vertical trend; on the other hand, it can be noted that, at the pier, the deepness algorithm shows more than one discontinuity, making it more difficult to identify the two actual sea/detached-breakwater crossings. For this reason it was decided to deepen the study of this algorithm using the coastal blue band.

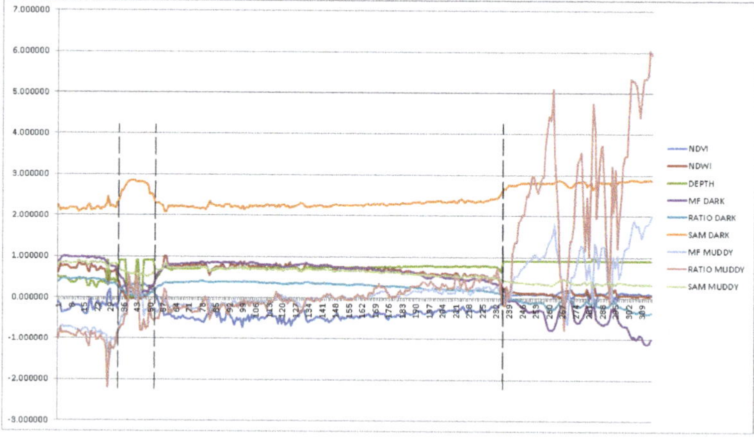

Figure 13. Comparison of the different algorithms on the same profile, black dashed lines are the actual water/ground limits (distances in the x direction are in meters along the profile, values in the y direction are reported for every specific algorithm so are not directly comparable).

4.3. The Coastal Blue Band and the Relative Depth Algorithm

The results obtained using the relative depth algorithm have suggested that it could be further enhanced to deepen the study by using the coastal blue band, although this is not present on all satellite platforms. One can get useful information not only on the coast but also on the first meters of the seabed thanks to its ability to penetrate clear water. For the study of the coastline, we have observed how the algorithm assumes a value of zero in the presence of land emersion (Figure 14) while reconstructing the profile of the seabed moving to the sea.

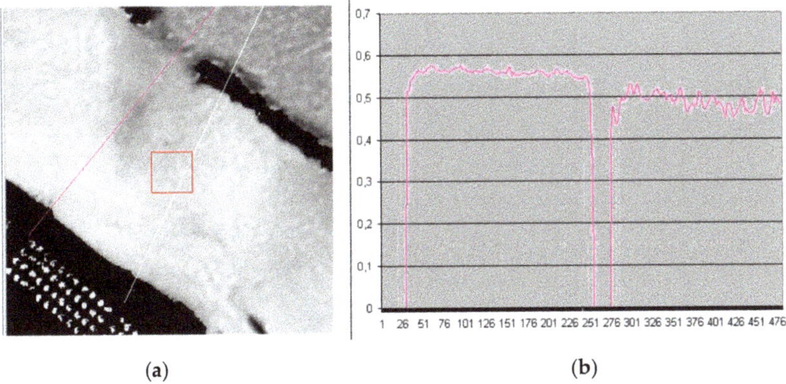

(a) (b)

Figure 14. Detail of relative depth algorithm results using coastal blue band; on (**a**) the profile trac, on (**b**) the resulting profile were open water is on the right, distances in the x direction are in meters along the profile.

However, the depth is relative, and therefore some known depths are necessary to transform the relative depths into absolute values. The only official data on the depths in this area are those released by the Military Geographic Institute (IGMI), collected with various techniques and accuracy. The correct calibration of the results of the algorithms would require more accurate measurements taken at the same time as the acquisition of the image. The data provided by the IGM only allow us to make a rough evaluation of the trends of the algorithm itself. In this way, we obtained the absolute depth profile (Figure 15).

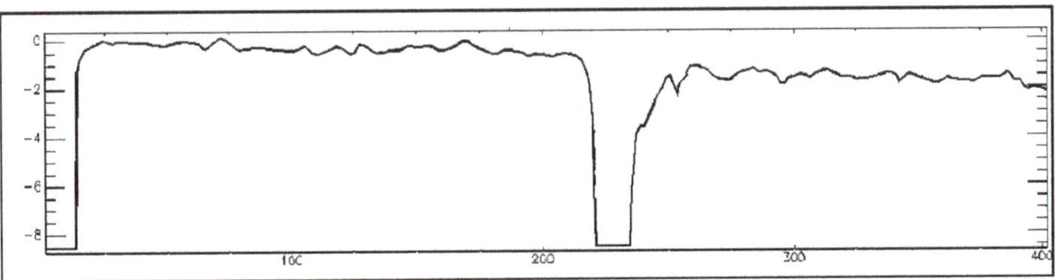

Figure 15. "Absolute depth" profile: along the *y* direction is the absolute depth and along the *x* direction is the progressive distance; all distances are in meters and open water is on the right.

The general trend is very similar to the expected one (Figure 16), also highlighting the differences in altitude near the detached breakwater. Despite the use of a median filter, some circular bathymetric curves are observed that might not respond to real morphologies. This could be due to the high detail of the information or to some bias of the algorithm; to be able to evaluate with certainty the performance of the algorithm, it would also be necessary to have a survey with greater accuracy (for example, side scan sonar) performed at the same time as the acquisition of the image.

Figure 16. "Absolute depth" bathymetric lines; open water is on the right.

4.4. Supervised Multispectral Classification

To complete the framework of the experiments, a supervised multispectral classification was also performed using the maximum likelihood algorithm; for this purpose, several regions of interest (ROI) were defined, including four building classes and three vegetation classes (Figure 17).

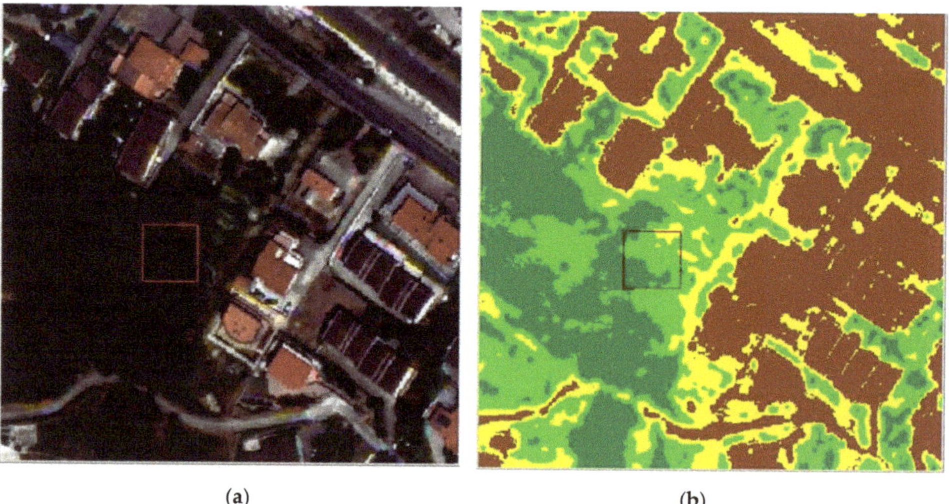

(a)   (b)

Figure 17. Detail of a vegetated area on RGB image on (**a**) and corresponding classification classes on (**b**).

The result obtained from such a classification was a new raster file in which, in each pixel, the digital number corresponding to the class of land cover to which the pixel itself has been associated is reported (Figure 18). This classification works equally well in the division of the various types of buildings and vegetation as well as with the separation of dry sand and wet sand. We compared the instantaneous shoreline thus obtained with the only official one available, that is the one reported in the regional map scale 1:5000, aware of the shifts due to different period, time, and phase of the waves, however, finding a good agreement that can only confirm that the coastline in this area has not undergone major changes.

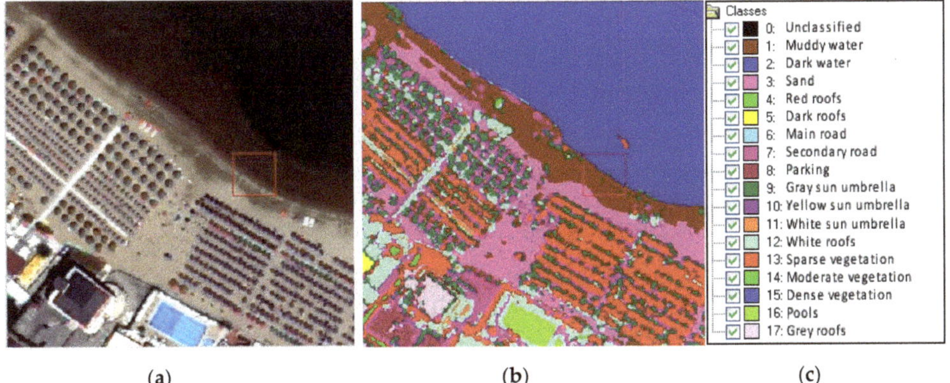

(a)   (b)   (c)

Figure 18. (**a**) Pansharpened image; (**b**) classified image; (**c**) classes legend.

5. Conclusions and Further Developments

The purpose of this project was to identify the most efficient procedure for automatically extracting the instantaneous shoreline in sandy coasts of the Abruzzo region that are affected by severe erosion problems. What was important to observe were the possible discontinuities in the digital numbers of the images after treatment with the various algorithms; in fact, once any discontinuities have been identified, the edge-detector functions are able to easily draw border vector lines on the same discontinuities. Therefore, different algorithms have been tested and their trends on the sea/beach passage have been examined, comparing them with the limits identified on the same image by an experienced photogrammetric operator.

The algorithm that showed a trend with more sharp discontinuities and coinciding with those identified by the operator was the relative depth. However, this algorithm showed more discontinuities than those observable on the image at the pier. For this reason, it was subsequently experimented with to see if it was possible to improve the results using the coastal blue band. It has been verified that the costal blue with the relative depth eliminates the problem of the multiple detections. It can be deduced that the most accurate results for this specific application are obtained with the relative depth algorithm using the coastal blue band of the WorldView 2 satellite.

As far as the use of the same algorithm for the evaluation of bathymetry is concerned, it is necessary to have precise, punctual measurements to calibrate it and measurements such as side scan sonar to verify it; both should be carried out at the same time as the acquisition of the image to be significant.

Author Contributions: Supervision, M.A., V.B., R.B. and F.R.; Methodology, M.A., V.B., R.B. and F.R.; Investigation, M.A., V.B., R.B. and F.R.; Writing—Review and editing, V.B., R.B. and F.R.

Funding: This research received no external funding.

Conflicts of Interest: The authors declare no conflict of interest.

References

1. Dolan, R.; Hayden, B.P.; May, P.; May, S. The reliability of shoreline change measurements from aerial photographs. *Shore Beach* **1980**, *48*, 22–29.
2. Cervino, R.; Ivaldi, R.; Surace, L. Il ruolo dell'Istituto idrografico della Marina nel monitoraggio costiero. In Proceedings of the 3rd Symposium Il Monitoraggio Costiero Mediterraneo: Problematiche e Tecniche di Misura, Livorno, Italy, 15–17 June 2010.
3. Boak, E.H.; Turner, I.L. Shoreline definition and detection: A review. *J. Coast. Res.* **2005**, *21*, 688–703. [CrossRef]
4. Aguilar, F.J.; Fernández, I.; Pérez, J.L.; López, A.; Aguilar, M.A.; Mozas, A.; Cardenal, J. Preliminary results on high accuracy estimation of shoreline change rate based on coastal elevation models. *Int. Arch. Photogramm. Remote Sens. Spat. Inf. Sci.* **2010**, *33*, 986–991.

5. Klein, M.; Lichter, M. Monitoring changes in shoreline position adjacent to the Hadera power station, Israel. *Appl. Geogr.* **2006**, *26*, 210–226. [CrossRef]

6. Progetto Preliminare Della Costa Teatina—Fenomeni Erosivi Della Fascia Costiera. Available online: http://www.provincia.chieti.it/flex/cm/pages/ServeAttachment.php/L/IT/D/2%252F4%252F4%252FD.d367c153fd7837fd72ca/P/BLOB%3AID%3D3926/E/pdf (accessed on 28 October 2019).

7. MATTM-Regioni. *Linee Guida per la Difesa Della Costa dai Fenomeni di Erosione e Dagli Effetti dei Cambiamenti Climatici. Versione 2018—Documento Elaborato dal Tavolo Nazionale sull'Erosione Costiera*; MATTM-Regioni con il Coordinamento Tecnico di ISPRA: Rome, Italy, 2018; 305p.

8. Klemas, V. Airborne Remote Sensing of Coastal Features and Processes: An Overview. *J. Coast. Res.* **2013**, *29*, 239–255. [CrossRef]

9. Costantino, D.; Angelini, M.G. Thermal monitoring using an ASTER image. *J. Appl. Remote Sens.* **2016**, *10*, 46031. [CrossRef]

10. Tarig, A. New Methods for Positional Quality Assessment and Change Analysis of Shoreline Features. Ph.D. Thesis, School of the Ohio State University, Columbus, OH, USA, 2003.

11. Ullman, S. The Interpretation of Structure from Motion, The Royal Society. University of Southampton. 1979. Available online: https://www.jstor.org/stable/77505 (accessed on 12 December 2019).

12. Troisi, S.; Baiocchi, V.; Del Pizzo, S.; Giannone, F. A prompt methodology to georeference complex hypogea environments. *Int. Arch. Photogramm. Remote Sens. Spat. Inf. Sci.* **2017**, *42*, 639–644. [CrossRef]

13. Pugliano, G.; Robustelli, U.; Di Luccio, D.; Mucerino, L.; Benassai, G.; Montella, R. Statistical Deviations in Shoreline Detection Obtained with Direct and Remote Observations. *J. Mar. Sci. Eng.* **2019**, *7*, 137. [CrossRef]

14. Palazzo, F.; Latini, D.; Baiocchi, V.; Del Frate, F.; Giannone, F.; Dominici, D.; Remondiere, S. An application of COSMO-Sky Med to coastal erosion studies. *Eur. J. Remote Sens.* **2012**, *45*, 361–370. [CrossRef]

15. Toure, S.; Diop, O.; Kpalma, K.; Maiga, A.S. Shoreline Detection using Optical Remote Sensing: A Review. *Isprs Int. J. Geo. Inf.* **2019**, *8*, 75. [CrossRef]

16. Dai, C.; Howat, I.; Larour, E.; Husby, E. Coastline extraction from repeat high resolution satellite imagery. *Remote Sens. Environ.* **2019**, *229*, 260–270. [CrossRef]

17. Dominici, D.; Zollini, S.; Alicandro, M.; Della Torre, F.; Buscema, P.M.; Baiocchi, V. High Resolution Satellite Images for Instantaneous Shoreline Extraction Using New Enhancement Algorithms. *Geosciences* **2019**, *9*, 123. [CrossRef]

18. WorldView-2 Satellite Sensor. Available online: https://www.satimagingcorp.com/satellite-sensors/worldview-2/ (accessed on 12 December 2019).

19. Available online: https://www.harrisgeospatial.com/Support/Maintenance-detail/ArtMID/13350/ArticleID/23403/Whats-New-ENVI174-55 (accessed on 12 December 2019).

20. Maglione, P.; Parente, C.; Vallario, A. Coastline extraction using high resolution WorldView-2 satellite imagery. *Eur. J. Remote Sens.* **2014**, *47*, 685–699. [CrossRef]

21. Baiocchi, V.; Bianchi, A.; Maddaluno, C.; Vidale, M. Pansharpening techniques to detect mass monument damaging in Iraq. *Int. Arch. Photogramm. Remote Sens. Spat. Inf. Sci.* **2017**, *XLII-5/W1*, 121–126. [CrossRef]

22. Parente, C.; Pepe, M. Influence of the weights in IHS and Brovey methods for pan-sharpening WorldView-3 satellite images. *Int. J. Eng. Technol.* **2017**, *6*, 71. [CrossRef]

23. Vivone, G.; Alparone, L.; Chanussot, J.; Dalla Mura, M.; Garzelli, G.; Licciardi, A.; Restaino, R.; Wald, L. A critical comparison among pansharpening algorithms. *IEEE Trans. Geosci. Remote Sens.* **2015**, *53*, 2565–2586. [CrossRef]

24. Maurer, T. How to pan-sharpen images using the gram-schmidt pan-sharpen metho—A recipe. *Int. Arch. Photogramm. Remote Sens. Spat. Inf. Sci.* **2013**, *XL-1/W1*, 239–244. [CrossRef]

25. Kruse, F.A.; Lefkoff, A.B.; Boardman, J.B.; Heidebrecht, K.B.; Shapiro, A.T.; Barloon, P.J.; Goetz, A.F.H. The Spectral Image Processing System (SIPS)—Interactive Visualization and Analysis of Imaging spectrometer Data. *Remote Sens. Environ.* **1993**, *44*, 145–163. [CrossRef]

26. Ferrier, G. Application of Imaging Spectrometer Data in Identifying Environmental Pollution Caused by Mining at Rodaquilar, Spain. *Remote Sens. Environ.* **1999**, *68*, 125–137. [CrossRef]

27. Stumpf, R.P.; Holderied, K.; Sinclair, M. Determination of water depth with high-resolution satellite imagery over variable bottom types. *Limnol. Oceanogr.* **2003**, *1*. [CrossRef]

Journal of
Marine Science and Engineering

Article

Benthic Habitat Morphodynamics-Using Remote Sensing to Quantify Storm-Induced Changes in Nearshore Bathymetry and Surface Sediment Texture at Assateague National Seashore

Arthur Trembanis [1,*], Alimjan Abla [2], Ken Haulsee [1] and Carter DuVal [1]

[1] School of Marine Science and Policy, University of Delaware, Lewes, DE 19958 USA; Khaulsee@udel.edu (K.H.); cduval@udel.edu (C.D.)

[2] Institute of Marine Science, Middle East Technical University, Ankara 06800, Turkey; alimjan_abla@yahoo.com

[*] Correspondence: art@udel.edu; Tel.: +1-302-831-2498

Received: 25 August 2019; Accepted: 16 October 2019; Published: 18 October 2019

Abstract: This study utilizes repeated geoacoustic mapping to quantify the morphodynamic response of the nearshore to storm-induced changes. The aim of this study was to quantitatively map the nearshore zone of Assateague Island National Seashore (ASIS) to determine what changes in bottom geomorphology and benthic habitats are attributable to storm events including hurricane Sandy and the passage of hurricane Joaquin. Specifically, (1) the entire domain of the National Parks Service offshore area was mapped with side-scan sonar and multibeam bathymetry at a resolution comparable to that of the existing pre-storm survey, (2) a subset of the benthic stations were resampled that represented all sediment strata previously identified, and (3) newly obtained data were compared to that from the pre-storm survey to determined changes that could be attributed to specific storms such as Sandy and Joaquin. Capturing event specific dynamics requires rapid response surveys in close temporal association of the before and after period. The time-lapse between the pre-storm surveys for Sandy and our study meant that only a time and storm integrated signature for that storm could be obtained whereas with hurricane Joaquin we could identify impacts to the habitat type and geomorphology more directly related to that particular storm. This storm impacts study provides for the National Park Service direct documentation of storm-related changes in sediments and marine habitats on multiple scales: From large scale, side-scan sonar maps and interpretation of acoustic bottom types, to characterize as fully as possible habitats from 1 to 10 m up to many kilometer scales, as well as from point benthic samples within each sediment stratum and these results can help guide management of the island resources.

Keywords: side-scan sonar; swath bathymetry; habitat monitoring; hurricane Sandy; hurricane Joaquin; climate change; shoreline detection; remote sensing

---

1. Introduction

The overall goal of this study was to map the morphodynamic changes to the nearshore zone of Assateague Island National Seashore (ASIS) to determine what changes in bottom sediments and benthic habitat occurred are attributable to storm impacts such as Superstorm Sandy and hurricane Joaquin. Specifically, we (1) mapped the full survey area with side-scan sonar and multibeam bathymetry at a resolution comparable to that of the pre-storm survey, (2) resampled a subset of the benthic stations that represented all sediment strata previously identified, and (3) compared newly obtained data to that from the pre-storm survey to examine storm impacts. This survey design aimed to document storm-related changes in sediments and marine habitats on multiple scales: From large scale, side-scan sonar maps and interpretation of acoustic bottom types, to characterize as fully as possible habitats from 1 to 10 m up to many kilometer scales, as well as from point benthic samples within each sediment stratum.

1.1. Background on Acoustic Mapping

Human reliance upon, and interference with benthic ecosystems necessitates an understanding of the spatial extent, structure, and function of these unique ecosystems [1,2]. This project work utilized advanced geoacoustic techniques for remote benthic habitat mapping and employed both traditional technologies for wide area coverage combined with point sampling and robotic systems including an autonomous underwater vehicle (AUV) and a remotely operated vehicle (ROV) for gathering data in an unprecedented level of resolution over targeted regions of interest. Knowledge gained from this study thus allows coastal and estuary environmental stakeholders responsible for the Assateague Island National Seashore (ASIS) to better manage resources to protect these environments by providing information on the location of habitats, species' distributions and associations to varying sedimentary and morphological regimes. The new benthic habitat and geomorphologic datasets will serve as a useful reference for an array of mapping and management efforts particularly providing insights into temporal dynamics from before and after storms.

Estuarine and coastal regions of the US face multiple marine spatial planning issues and challenges, including the effects of episodic storms, climate change, and other stressors [3]. In particular, there has been a strong interest in and multiple efforts to examine the effects of hurricane Sandy on the seabed within the Mid-Atlantic Bight [4–6]. The anticipated impacts of offshore development on the seabed and to the ecology and, more generally, initiatives in marine spatial planning [3] point to an ever-increasing need for both base-line mapping and monitoring efforts directed at biological habitats and geological features of the littoral zone.

In this study, we realized the need to fuse disparate data streams and recognized the most effective means of analyzing and understanding them is through data visualization. By repeated benthic habitat and morphology mapping, this field effort and the subsequent data analysis provides a critical extension to the previous study of the ASIS site (2011–2012). We recognize that together these data sets collected over multiple years represent a knowledge base much greater than the sum of its parts, and it is this fact that motivated the project.

Recent benthic mapping on similar prior projects to this project [7–9] have demonstrated that, in general, shallow coastal ecosystems are far more complex and heterogeneous in bottom type than typically anticipated. Furthermore, conventional sampling by direct methods (i.e., bottom grabs or dredges) is also generally recognized as severely limited by several factors including gear bias, spatial averaging and limited sample density. It is only from the careful fusion of multiple technologies (i.e., ship-based, ROV, AUV, grab samples, and dredge) that one is able to compile the complete picture of the nature and composition of the bottom [8–11].

Accurate bathymetric and backscatter data are essential for hydrographic charts, dredging, and habitat mapping. Previously, the primary method for seafloor mapping was the use of single-beam bathymetry. It is an accurate technique for collecting seafloor data, yet there are a number of limitations. In single-beam surveys, a discrete footprint of the seabed is ensonified leaving large gaps between adjacent survey tracks that are not surveyed. Single-beam surveys typically cover only 5-10% of the total seafloor area leaving large portions unsampled. Therefore, the maps generated from single-beam sonar samples may lack high-enough resolution to detect small and ecologically important habitat features [12]. In a recent study [12], the authors found that a full coverage swath sonar system resulted in a 90% classification accuracy compared to 70–80% accuracy from a single beam system approach. Using conventional methods like RoxAnn (e.g., [13,14]), found that the sampling is limited to a single sonar point footprint directly below the vessel, and maps necessitate heavy interpolation between track lines. In shallow settings, single-beam systems therefore require progressively more and more tightly spaced lines as the local depth decreases. In contrast, multi-beam echosounders, interferometric sonar, and side-scan sonar systems characterize a wide swath on either side of the survey vessel. Surfaced towed acoustic sensors suffer from positioning errors associated with the ambiguity of the towfish location relative to the ship global positioning system (GPS). Surface vessel mounted sensors also suffer from reduced resolution (larger footprint) as water depth increases. Wide swath acoustic methods afford the potential of effectively mapping large areas of the seafloor at a resolution appropriate for habitat mapping and correlation with biological communities. Seafloor echoes can be interpreted in terms of hardness and roughness and correlated with the sediment and faunal characteristics.

Several reviews and studies of coastal sedimentary systems (e.g., [15–17]) have repeatedly demonstrated that dramatic spatiotemporal heterogeneity dominates seabed settings with respect to composition and morphology and that this heterogeneity is more the norm than the exception. Further exacerbating the problem, seabed heterogeneity often exists at spatial scales on the order of 1–10's m, well below typical single-beam survey spacing and gridding schemes. Understanding and being able to map this spatial patchiness at sufficient coverage and resolution scales is a critical component of habitat mapping efforts and an important motivation behind this project.

1.2. Study Design and Objectives

This study was designed to characterize the physical and biological components of the nearshore benthic zone of Assateague Island National Seashore (ASIS), including delineations of habitat types using the CMECS system. We characterized benthic sedimentary habitats by analyzing Young grab samples for grain size distribution, and total organic carbon (TOC). The study methods were modeled off of the 2011 ASIS benthic habitat assessment projects undertaken by Versar, Inc. (Columbia, MD) and the Maryland Geological Survey (MGS), originally contracted by the National Park Service (NPS). A Natural Resource Condition Assessment undertaken by ASIS and the University of Maryland [18] ranked the Atlantic subtidal habitats as reaching 99% of their reference condition (as determined by historical observations and data), albeit with very limited confidence or knowledge of indicators for the reference condition. Thus, data from the assessments by Versar, Inc. and the Maryland Geological Survey serve as a crucial pre-Sandy (2011) point of comparison for the possible effects of the physical stressors of hurricane Sandy on sediment distribution. While this project was one of four simultaneous benthic mapping regimes funded by the NPS in the aftermath of Sandy, this study covered the largest habitat area, focused on the Atlantic subtidal zone, and was unique in its availability of pre-Sandy (2011) reference data.

In this paper we set out to: (1) Provide an updated subtidal resource inventory to the National Park Service by determining benthic distribution utilizing geoacoustic surveys compelled with grab samples for grain size analysis; (2) analyze pre-Sandy (2011) and post-Sandy (2014-2015) habitats with a uniform statistical approach and compared them to determine what changes occurred through the study period; (3) conduct targeted acoustic mapping of areas of particular interest before and after storms to provide an unprecedented higher resolution of these target locations.

1.3. Study Site

Assateague Island is a 58-km long barrier island stretching from the Ocean City inlet in Maryland, down past Chincoteague Island in northern Virginia. This dynamic island is characterized by barrier beach and back-bay environments that are subject to the constant eroding and re-shaping forces of the Atlantic Ocean. The pre-storm and post-storm surveys of Assateague Island subtidal resources discussed here include the majority of the length of the island (Figure 1). We sampled benthic sediments up to 1 km offshore in accordance with the pre-storm survey and NPS jurisdictions (Figure 2), whereas the USGS surveyed areas further offshore of Assateague Island. Historically the nearshore subtidal habitats at ASIS are comprised mostly of fine sand, with patches and swaths of mud and coarser sediments. Much of the observed heterogeneity in sediments is due to the relict sediments left behind by the characteristic landward 'rolling-over' of barrier islands.

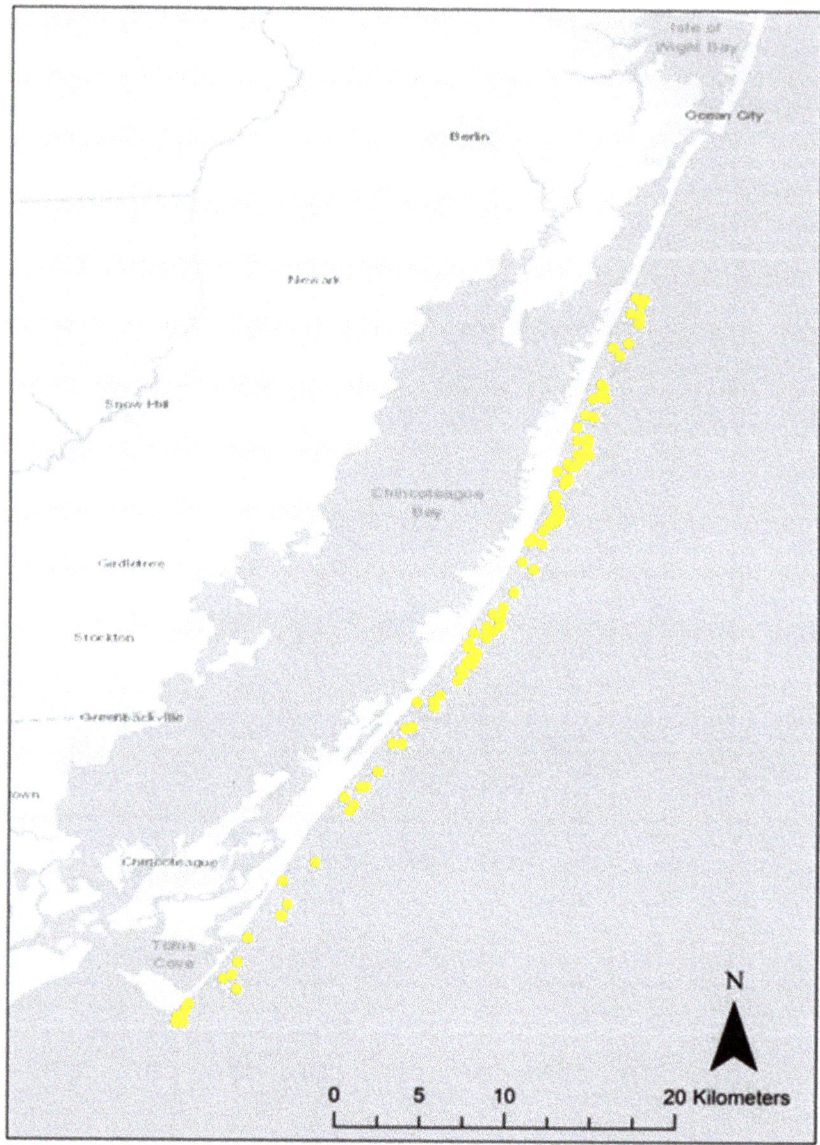

Figure 1. Pre-Sandy (2011) survey sites (n = 125) sampled in October 2011 on the beach (n = 15) and subtidally up to 1 km offshore (n = 110) of Assateague Island National Seashore, Maryland.

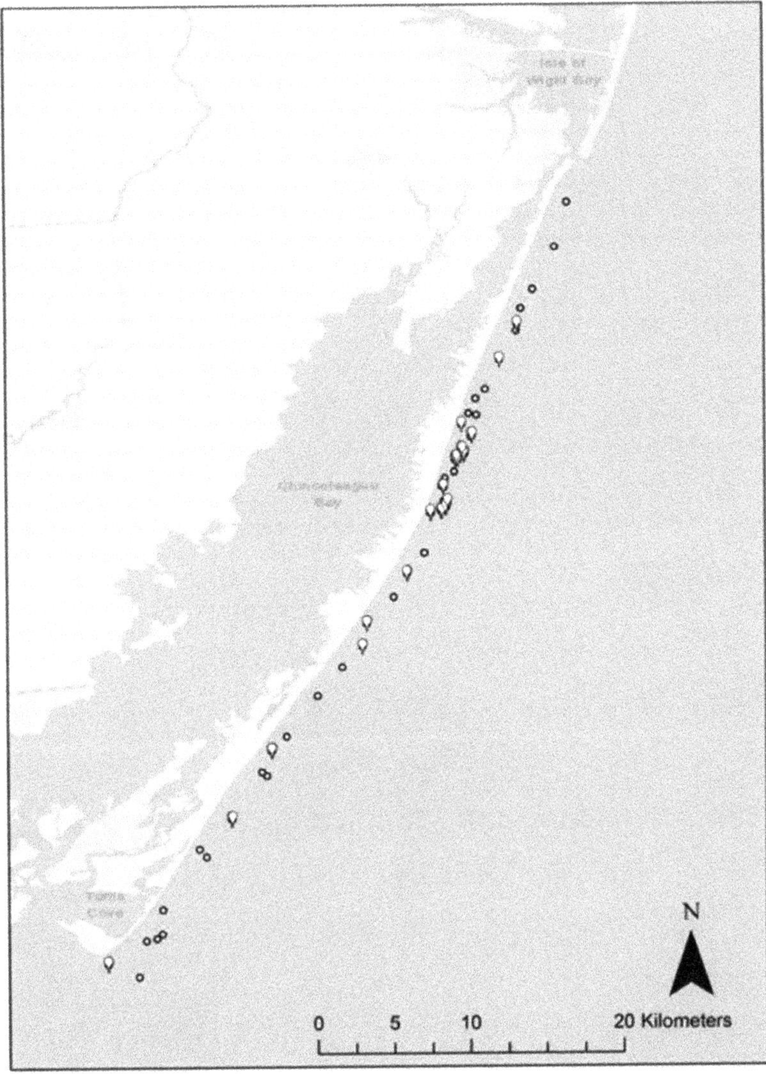

Figure 2. Post-storm (n = 120 grabs) sampled in October 2014 (n = 60 grabs, white) and 2015 (n = 60 grabs, yellow), up to 1 km offshore of Assateague Island National Seashore survey sites, Maryland.

2. Materials and Methods

2.1. Acoustic Mapping: Side-Scan Sonar Workflow

The side-scan sonar data collected was processed using Chesapeake Technology, Inc., Mountain View, CA, USA SonarWiz 6.0 sonar processing software. Sonar data processing followed industry standard procedures for side-scan data, focusing on gain corrections to remove the variation in across track brightness inherent in side-scan sonar performance in order to achieve the most representative mosaic of the seafloor. Standardized gain settings were chosen to make the sonar data as internally consistent as possible between different daily mission datasets, while also attempting to replicate the products produced from the Versar sonar data collected off Assateague in 2011. Raw Edgetech sonar files (.jsf) were imported using a time-varying gain, to compensate for signal loss on the outer swath of the sonar files. Once the files were imported, bottom-tracking corrections were applied if automatic bottom tracking had failed. Once verified or corrected, an empirical gain normalization correction was applied. This gain correction makes the data set consistent from a file-to-file basis, removing gain biases caused by variations in backscatter intensity from differing sediments between each file. This produces an interiorly consistent mosaic. From this work flow (Figure 3), several mapping products were created. A standard georeferenced image file (geotiff) and Google Earth georeferenced file (.kmz) are generated. Vessel navigation or track-lines files were exported as shapefiles (.shp). Each gain-corrected file was exported as a waterfall image (.jpg). Target reports were generated (in .pdf format), highlighting unique features or objects on the seafloor. Finally, coverage reports (.xls) list the metadata for each file (start/end latitude and longitude, length, time, etc.), as well as the total area and line-line overlap achieved for the complete mission.

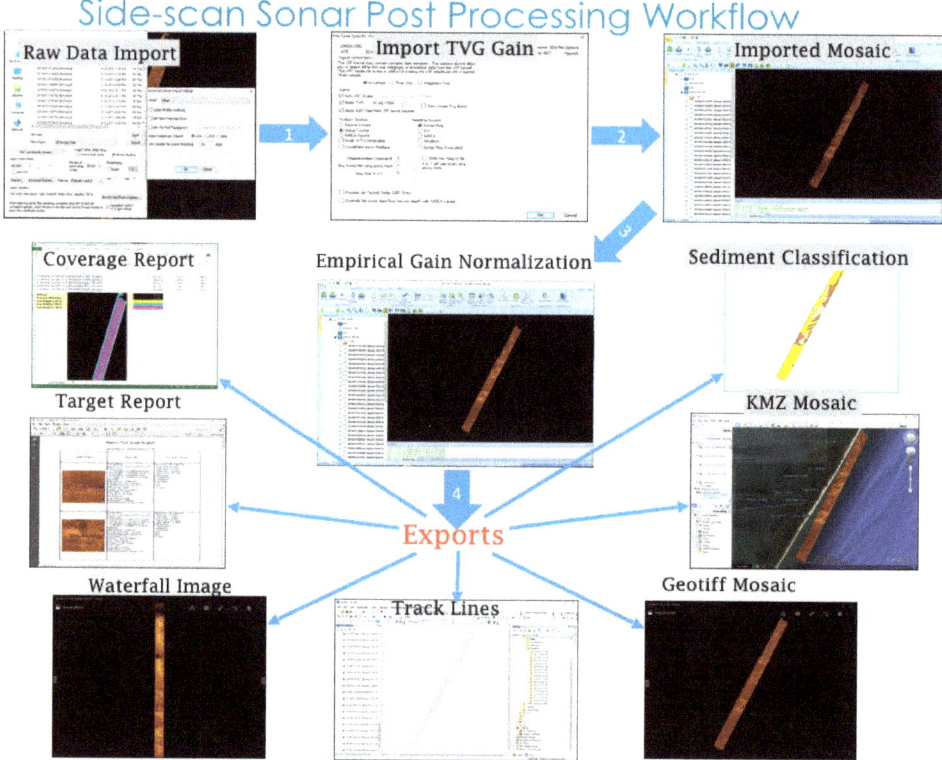

Figure 3. Side-scan sonar post processing workflow used by the University of Delaware, Assateague Island National Seashore mapping project: From EdgeTech raw data to SonarWiz to shapefiles and Google Earth .kml data products.

2.2. Bathy Workflow

The bathymetric data collected was processed using Chesapeake Technology, Inc., Mountain View, CA, USA SonarWiz 6.0 sonar processing software. Sonar data processing followed industry standard procedures for bathymetric data, focusing on injecting tidal offsets, sound velocity offsets, and configuring the vessel files in order to achieve the most representative mosaic of the seafloor. The bathymetric processing workflow (Figure 4) was standardized to keep the bathymetric representations as internally consistent as possible. The first step was to configure a vessel offset file for the vessel that was recording the sonar data. All of the research vessels measurements were captured including the inertial measurement unit (I.M.U.) to the echosounder and the I.M.U to the antennas were recorded and saved in a Sonarwiz6 vessel file. After the vessel file is created, the raw sonar data (.jsf) was imported into a new Sonarwiz6 project that was set-up with an internally consistent naming convention based off the date of the survey. Before the data was merged, the sound velocity profiles, tide file, and the patch test values were imported into the project. After the echosounder was installed and before surveying operations began, a patch test was completed to ensure that each installation of the echosounder provided internally consistent data. The process accounts for offsets in roll, pitch, time latency, and heading. These patch test values were then entered into the vessel offsets file. In the field, sound velocity profiles were collected every hour on a Castaway CTD. The sound velocity files (.csv) were then imported into the Sonarwiz6 project. A tide file from the days that we surveyed was downloaded from the National Oceanic and Atmospheric Administration (NOAA) Tides and Current website in the form of .csv and then imported into the Sonarwiz6 project. After the patch test values, sound velocity profiles, and the tide file were imported into the project, the data was merged. This merging process injects all of the above information onto the XYZ point cloud that was recorded during the survey to constrain the positional error, while providing the most replicable data products that are internally consistent. Gridded mosaics, at a resolution of 0.75 m, were then exported as a .grd file using the CUBE algorithm, which is now the industry standard for providing the most accurate gridded solutions. The backscatter from the bathy was also processed in Sonarwiz6 and was exported as a .grd file. In all, a geotiff, .kmz, .xyz, and .grd of both the bathymetry data and the backscatter data were exported as bathymetric data products.

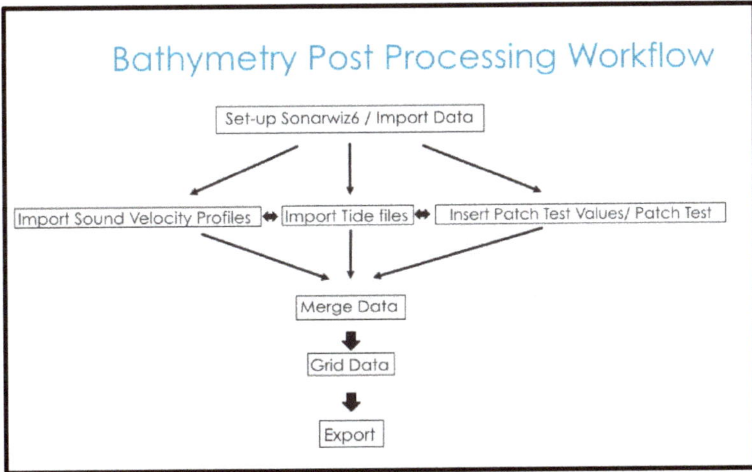

Figure 4. Schematic of workflow used by the University of Delaware mapping team for multibeam bathymetry data from EdgeTech 6205 acquisition to gridded data products. Data collected as part of a submerged mapping study for Assateague Island National Seashore in 2014–2015.

Based on the accuracy and resolution of our navigation system (a Coda F190R with dual head RTK and OmniSTAR enabled GNSS receivers) and standard patch test performed during our survey along with confirmation from overlapping passes we estimate that the horizontal and vertical uncertainty is approximately 25–30 cm respectively. Previous studies conducted using this same positioning system, sonar, and vessel in a further offshore site with the exact same sensors and configurations yielded repeatable target acquisition of within 1 m horizontally and was reported by [19]. Another study by the U.S. Geologic Survey (USGS) utilizing the same inertial measurement unit (IMU) and satellite correction service also yielded similar uncertainty values of general less than 50 cm for 95% of the hydrographic soundings [20]. Therefore, it is reasonable to set our achievable positioning reliability and overall total propagated error at approximately 50 cm horizontally and vertically. Our baseline inertial measurement accuracy was approximately 5cm horizontal and vertical and was aided by either OmniSTAR corrections or RTK base stations when available. Even without the satellite or RTK corrections our positioning would have been accuracy to a meter or two and thus much less than the scale of the changes seen across the seabed that range from 20 to 180 m of variability.

2.3. Acoustic Seabed Classification Map Preparation

2.3.1. Repeated Seabed Mapping Surveys

Our three, 2014–2015 acoustic surveys allowed us to make both inter and intra-annual comparisons for storm-related changes in bottom sediments. The first survey resulted in side-scan sonar mosaics and sediment acoustic class shapefiles from June 2014. These data include 5 side-scan sonar mosaics and sediment class shapefiles that were generated from the side-scan sonar data collected during 20–25 June 2014. The second survey was the most broadly defined and resulted in side-scan sonar mosaics and sediment class shapefiles for May 2015. These data include 12 side-scan sonar mosaics and sediment class shapefiles that were generated from the side-scan sonar data collected during 12–21 May 2015.

Finally, we conducted a set of spot selected surveys in October 2015 producing side-scan sonar mosaics and sediment class shapefiles. This survey was spatially focused on areas of change following hurricane Joaquin. Data were collected during 13–16 October 2015.

2.3.2. Acoustic Seabed Classification

First, all the available datasets were imported into ArcGIS platform under coordinate system - WGS_1984_UTM_Zone_18N (EPSG:32618). Then, manually digitized acoustic classes were matched against side-scan sonar mosaics and grain size data of grab samples to make sure the side-scan sonar mosaics were segmented correctly. After that, sediment class boundaries were matched between neighboring sediment classes to correct any inconsistencies in the interpretation of side-scan sonar mosaics, (i.e. to keep the class assignments consistent from one survey day to the next). Minor discrepancies with the side-scan sonar data interpretation and inconsistencies in sediment class boundaries were corrected manually using ArcGIS Edit tool.

After fixing all the discrepancies related with side-scan sonar data interpretation, all the same sediment classes in different shapefiles were merged into a single shapefile using the "Merge" tool in Arc Toolbox (Figure 5). Overlapping areas of the same sediment classes between different shapefiles were dissolved into one using the "Dissolve" tool in Arc Toolbox. At the end, 4 single shapefiles representing 4 different sediment classes namely Coarse Sand, Medium Sand, Fine Sand, and Mud, were merged into one single shapefile named Sediment Class_201406_201505.shp. A new field named "Area" was added to the table of contents of the newly generated shapefile and areas in m^2 were calculated for the sediment classes using "Calculate Geometry" method in ArcGIS.

The acoustic backscatter data collected in the field surveys was used to generate mosaic sonar images as outlined in the section above and then digitized using ArcGIS to manually outline and build categorical maps of the Assateague Island survey domain. The same acoustic class types as used in the 2011 precursor study by Versar/MGS were adopted here for consistency and the previous acoustic class maps were used along with the newly acquired and created side-scan sonar mosaics to guide the human operator in the segmentation process. All segmentation effort was conducted by the same team member for consistency and then reviewed by the principle investigator. The backscatter segments were then assigned to the acoustic class and were assigned similar colors again adopting the same nomenclature and color scheme as used in the precursor study to aid in comparison. Note that up to this point the classes are simply acoustic classes and require some external ground-truthing to further relate the classes to actual seabed types. The next step of relating the acoustic classes to seabed types comes through the correlation of class maps to the benthic grab samples and grain size analysis samples.

Figures 3 and 4 illustrate the workflows used in data processing both the side-scan sonar and phase measuring bathymetric sonar.

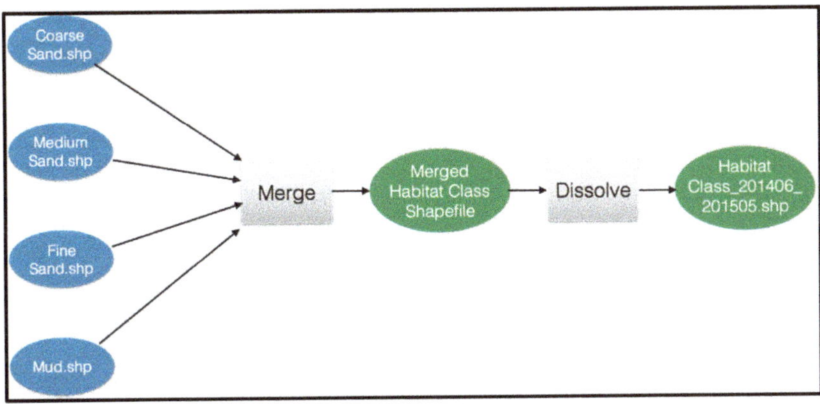

Figure 5. Flow-chart of data processing steps for manually segmented side-scan sonar-sediment classification into a habitat map for Assateague Island National Seashore, developed by the University of Delaware 2014–2015.

2.3.3. Ground Truthing

Essential for any geoacoustic marine habitat mapping initiative is the need to gather ground truthing observations to compare to the acoustically derived maps. Throughout our previous habitat mapping campaigns we have utilized benthic grab samples for ground truthing [6,9]. Grab samples were analyzed for grain size distribution and benthic faunal composition using standard approaches (see below and references therein for details). We used a modified Young grab sampler for recovering surface sediment ground truthing samples. The available grain size data included grain size analysis results from 98 grab samples taken during June of 2011 and 146 grab samples taken during October of 2014 and October of 2015 for benthic samples plus additional ones for ground truthing only. However, only the ones taken during 2014 and 2015 were used to check the side-scan sonar data interpretation in this study. Additionally, a small portable YSI Castaway CTD (http://www.ysi.com/castaway) was used to gather vertical profiles of salinity, temperature, and sound speed for use in environmental characterization and swath bathymetry data processing.

3. Results

3.1. Acoustic Mapping Products and Findings?

In total, over 65 km^2 of geoacoustic mapping was accomplished along the entire length of Assateague Island from as near to the shore as was possible out to at least 1km offshore. Additional sonar lines were acquired in order to connect the nearshore ASIS park boundary to the regional USGS mapping effort (Figure 6). The mapping area covered depths from 2 m–12 m extending from the beach to ~2 nm offshore, with final map resolution 50 cm/pixel for side-scan sonar and 1 m^2/pixel for bathymetry products. The field survey design as developed and agreed to by all project partners conducting mapping at this and other post-Sandy NPS sites was to achieve 100% side-scan sonar coverage and as much bathymetric coverage as possible given the constraints imposed by such shallow water survey operations. In this study we were able to achieve approximately 50% swath bathy coverage (Figure 7) and 100% side-scan sonar coverage (Figure 8).

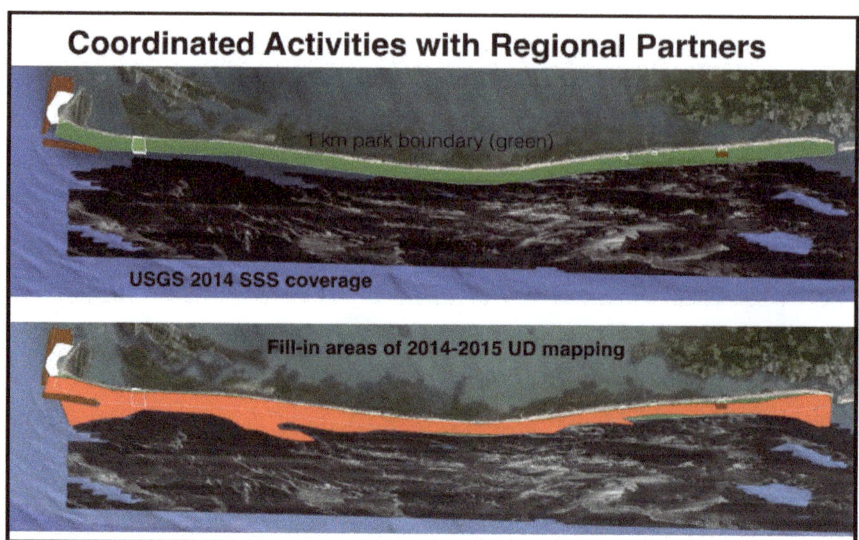

Figure 6. Overview of study site area showing park boundary (green) and recent USGS coverage along with UD mapping team gap filling coverage (red).

Figure 7. Bathymetry mosaic (1m/pixel) from 2014–2015 acoustic surveys (550 kHz) for the entire study area. Notable features captured in the bathymetry included pronounced shore-attached oblique bars.

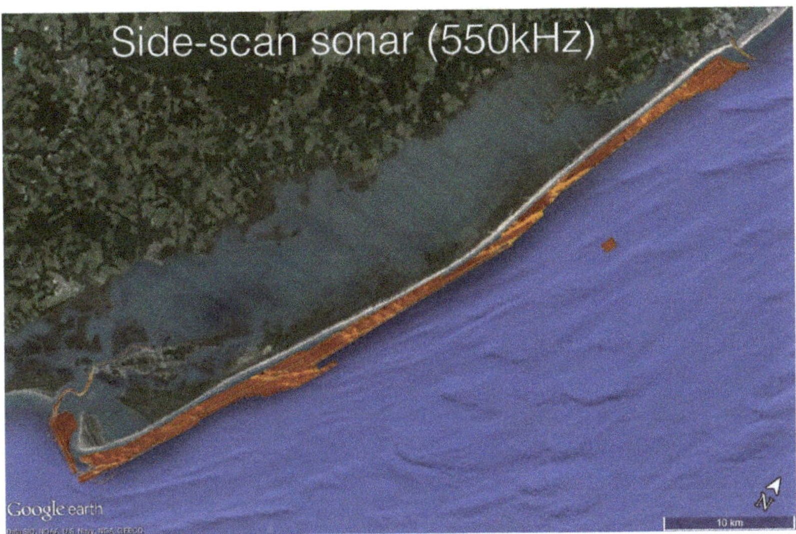

Figure 8. Side-scan mosaic from 2014–2015 acoustic surveys for the entire study area.

Resulting swath bathymetry data for the study side was gridded to a resolution of 1m/pixel. Reliable bathymetry data was obtained for approximately half the total sonar swath width resulting in 50% seafloor coverage in depths ranging from 2 to 12 m. Early sampling in 2014 around Fishing Point was conducted using spacing resulting in 100% bathymetry coverage (200% side-scan sonar) but the survey rate was unsustainable for the entire study site and it was agreed upon by all project partners to expand the line spacing to achieve 100% side-scan and 50% bathymetry. This survey strategy is also in keeping with the same approach used by the USGS in their larger regional coverage provided for a meaningful connection of this study to that regional effort. The resulting swath bathymetric grids extend what was only single-beam echosounder data from the precursor 2011 project and thus provide an enhanced baseline for AINS resource managers. Notable features captured in the bathymetry included pronounced shore-attached oblique bars.

A complete set of side-scan sonar mosaics with a resolution of 50 cm/pixel was generated for the entire length of the island and included extensions to fill gaps between the ASIS park boundary and the USGS regional survey effort. Gain settings and post-processing were set to generate an internally consistent sonar backscatter map (Figure 8) that allowed for consistent segmentation and interpretation with respect to sediment class. The convention used in the backscatter mosaics is for bright yellow color for high-amplitude returns (i.e. coarse sand and gravel) and dark brown/black for low amplitude returns (i.e. mud). Bright high-amplitude returns in the backscatter mosaic were clearly associated with the raised features of shore-attached oblique sand ridges noted in the bathymetry (Figure 7). Mud patches were most notable in the adjoining troughs associated with the sand ridges or from relict marsh outcrops exposed by scour and barrier island migration.

As is shown in Figure 9, the sediment classification map includes four different sediment classes were recognized and shaded with different colors.

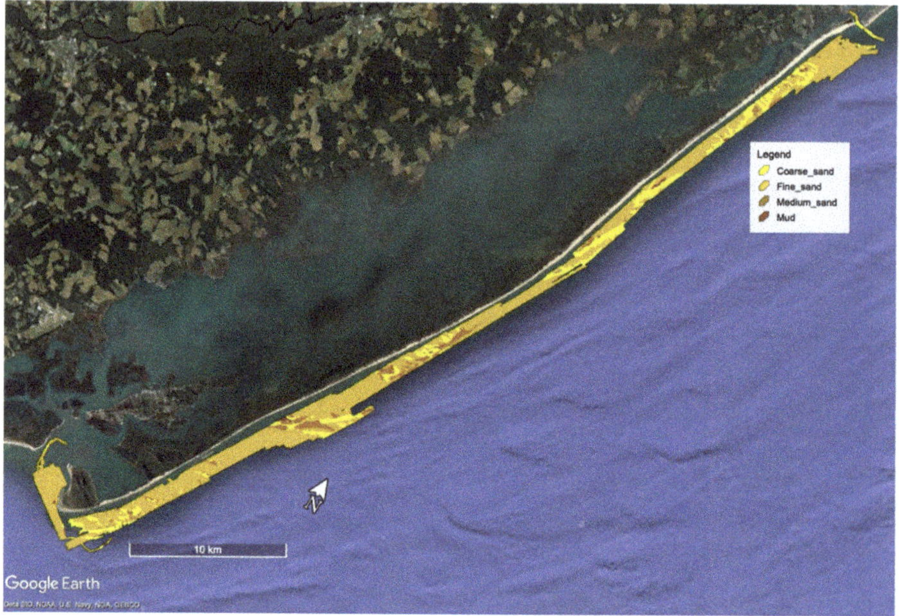

Figure 9. Sediment classification from 2014–2015 for the entire study area. Four sediment classes (coarse and medium sand, mud, and fine sand) were recognized and shaded with different colors. Mapping survey conducted 2014–2015.

Segmentation and acoustic classification of the side-scan sonar backscatter mosaic was performed via manual heads-up segmentation utilizing the same class types from the 2011 precursor study but informed by the sediment grab samples collected in this study. The GIS shapefile vectors for each acoustic class type are shown in Figure 9 using the same color and naming convention from 2011 for ease of comparison. The distribution of class type (Table 1) followed similar overall trends as documented in 2011 with medium and coarse sands associated with positive relief features like the large shore-attached oblique ridges and mud associated with the adjoining troughs. Notably absent in the 2014–2015 data were any mappable patches of gravel. The percentage area distribution of each class type are summarized in Table 1.

Table 1. Surficial sediment distribution by class type based on the 2014–2015 habitat mapping expedition.

Sediment Type	Classification Map Coverage %
Coarse Sand	10%
Medium Sand	1%
Fine Sand	82%
Mud	7%

The following three figures illustrate an example of an area with pronounced spatiotemporal changes in acoustic class texture between the 2011 precursor map (Figure 10) and repeated surveys with the EdgeTech 6205 in 2014 (Figure 11) and 2015 (Figure 12). The same backscatter amplitude convention is used but the gain settings and sonar frequencies are slightly different between the 2011 and 2014/15 surveys. Nevertheless, the general trend shows a migration of a set of coarse sand patches and the burial of a previously exposed mud patch.

Figure 10. Side-scan mosaic and classification shapefiles based on 2011 survey data.

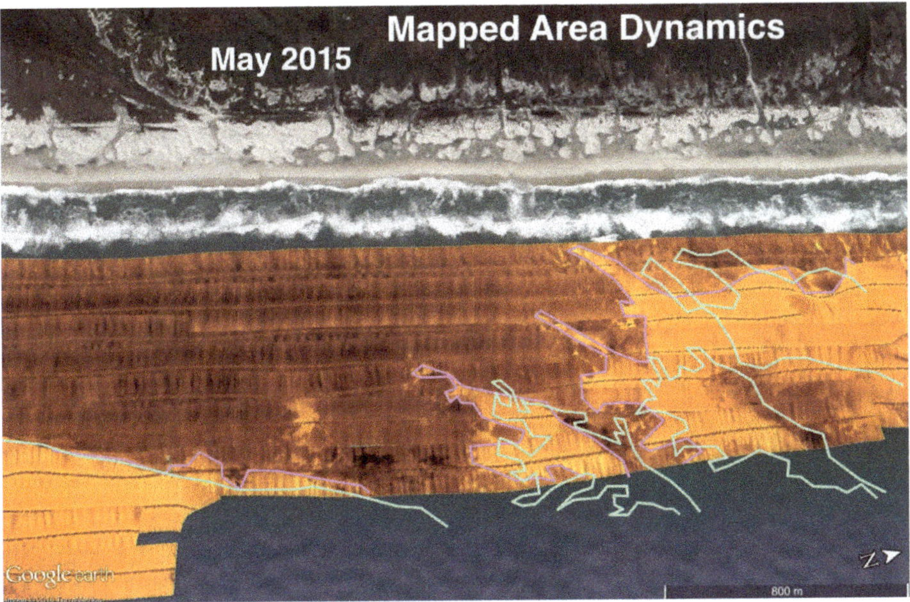

Figure 11. May 2015 side-scan mosaic and sediment classification shapefile compared to 2011 from Figure 9.

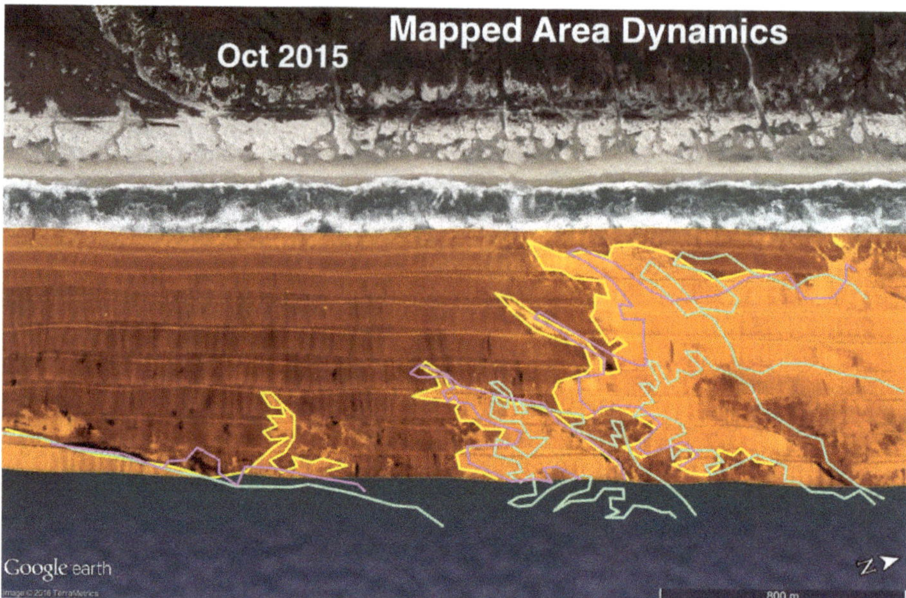

Figure 12. October 2015 side-scan mosaic and sediment classification shapefile compared to May 2015 (Figure 11) and 2011 (Figure 10).

3.2. Temporal Changes in Seafloor Sediment Type

Select portions of the survey domain were mapped both in May 2015 and October 2015 capturing before and after the passage of hurricane Joaquin. Sediment class maps were determined in the same manner and using the same grain size informed class types in order to examine changes in the before and after storm surveys. Sediment types were coded from 1 to 4 from smallest to largest grain size with 1 as mud, 2 as fine sand, 3 as medium sand and 4 as coarse sand and these codes were manually inserted into the table of contents of sediment class shapefiles. Sediment class shapefiles were then converted into raster files with sediment class type code being the value field. In order to understand if the seafloor sediments are coarsening or fining from May 2015 to October 2015, a sediment property change map was generated by subtracting sediment class raster data of May 2015 from sediment class raster data of October 2015 (Figure 13). This map shows the pattern and intensity of change in sediment properties from May 2015 to October 2015 as a magnitude of sediment property change index (Figure 13). For example, positive sediment property change index means the seabed became coarser, negative sediment property change index means the seabed became finer and zero value means sediment property was unchanged. Higher absolute value means higher intensity of change, for example +3 area represents a change from mud to coarse sand −3 area represents a change from coarse sand to mud.

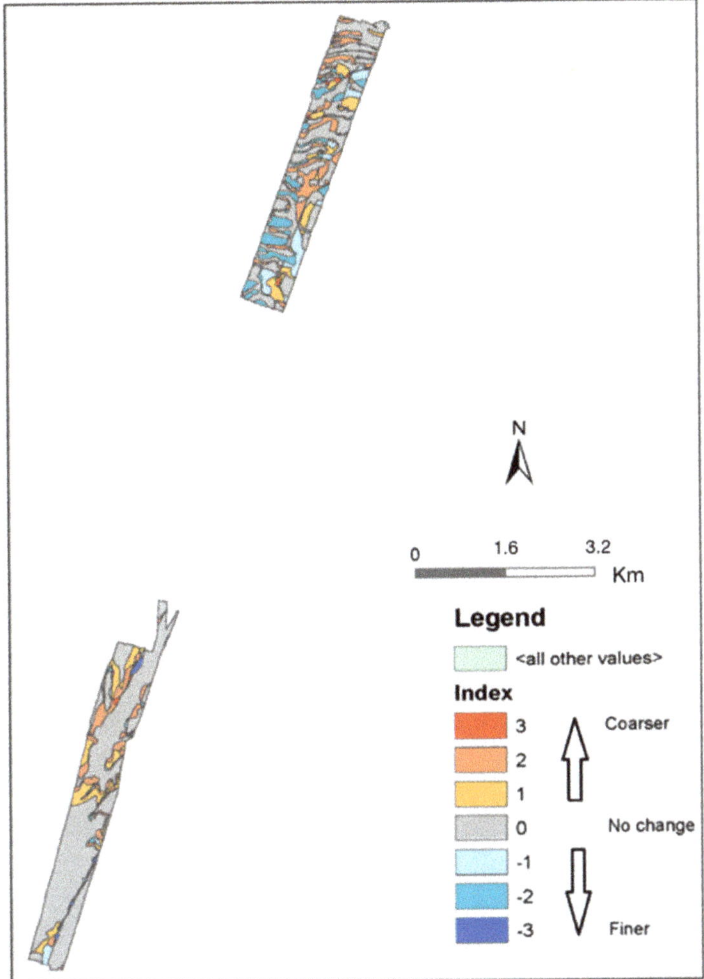

Figure 13. Sediment property change map for submerged resources of Assateague Island National Seashore between May 2015 and October 2015. Positive sediment property change index means seafloor became coarser, negative sediment property change index means seafloor became finer and zero value means sediment texture did not change.

Comparison of sediment class boundaries between May 2015 and October 2015 suggest a major shift (mostly south to eastward) in different sediment class boundaries. The distance of the shift in the fine-coarse sediment class boundary is about 20–40 m in the northern portion of the study area (Figures 14 and 15) and up to 80 m in the southern portion (Figures 16–19). The shift in mud-coarse sediment class boundary however, was measured to be up to 180 m in the southern part (Figure 17).

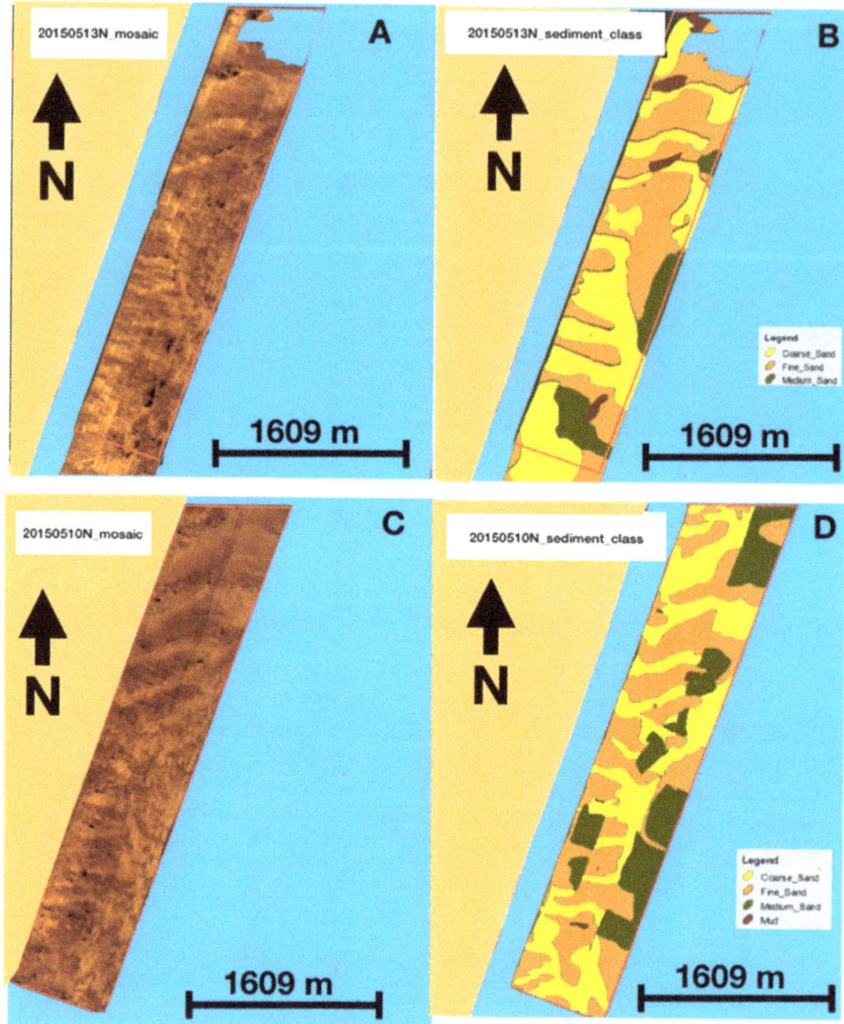

Figure 14. Side-scan sonar mosaic interpretation, Assateague Island National Seashore, Maryland (**A**) side-scan sonar mosaics collected during 13 May 2015; (**B**) interpretation of side-scan sonar data collected in 13 May 2015; (**C**) side-scan sonar mosaics collected in October 2015; (**D**) interpretation of side-scan sonar mosaics collected in October 2015. Note that red polygons represent the same areas mapped during both 13 May 2015 and October 2015.

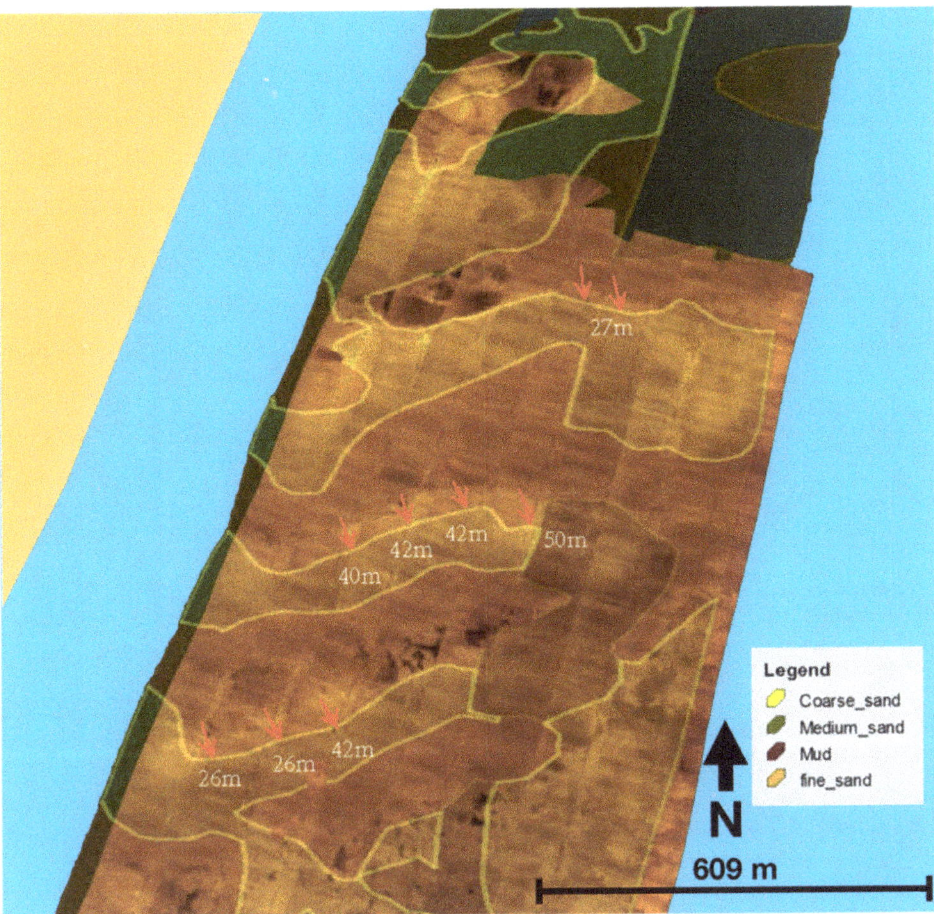

Figure 15. Temporal changes in sediment class boundaries, Assateague Island National Seashore, Maryland. The underlying image on the bottom is the side-scan sonar mosaic collected during May 2015. Shaded polygons on the top represent sediment classes inferred from the side-scan sonar mosaics collected during October 2015. Note the southeastward shift (18–50 m) in the boundary of fine-coarse sand sediment classes.

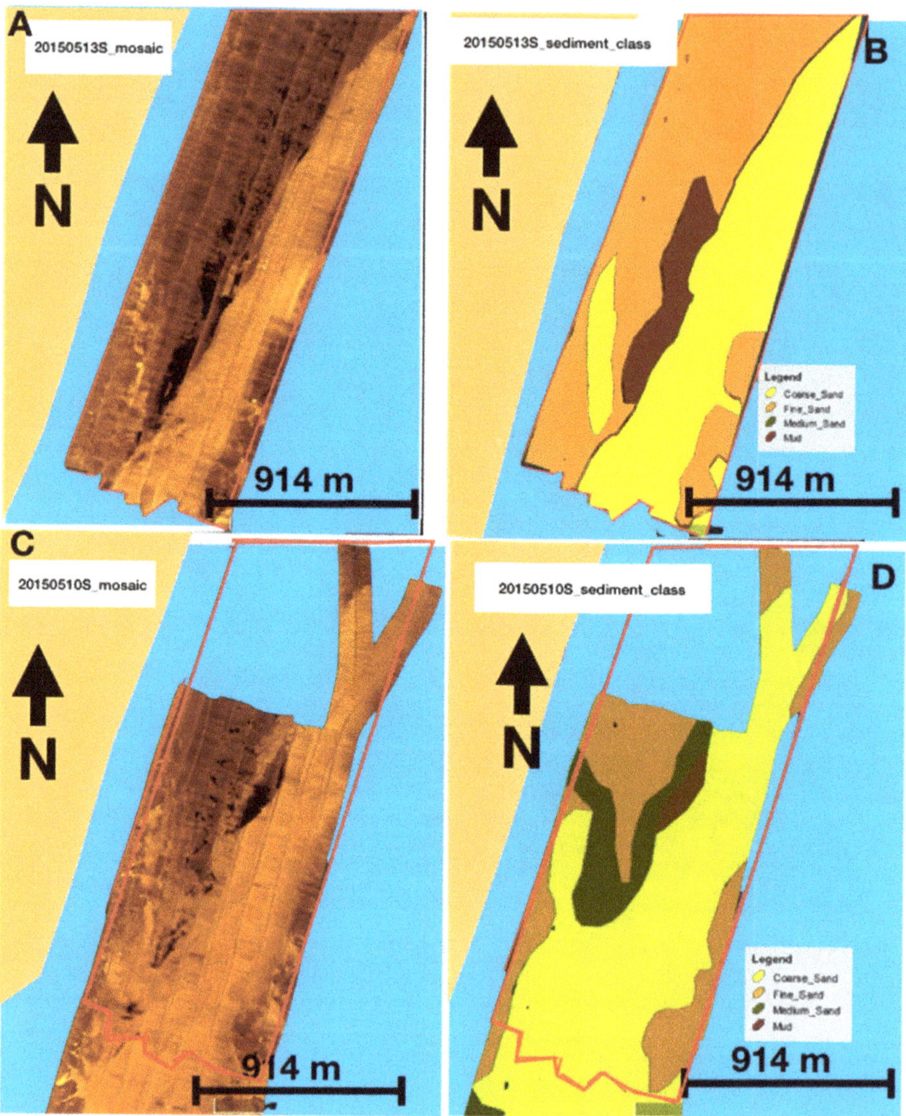

Figure 16. Side-scan sonar mosaic interpretation, Assateague Island National Seashore, Maryland. (**A**) Side-scan sonar mosaics collected during 13 May 2015; (**B**) Interpretation of side-scan sonar data collected in 13 May 2015; (**C**) side-scan sonar mosaics collected in October 2015; (**D**) Interpretation of side-scan sonar mosaics collected in October 2015. Note that red polygons represent the same areas mapped during both 13 May 2015 and October 2015.

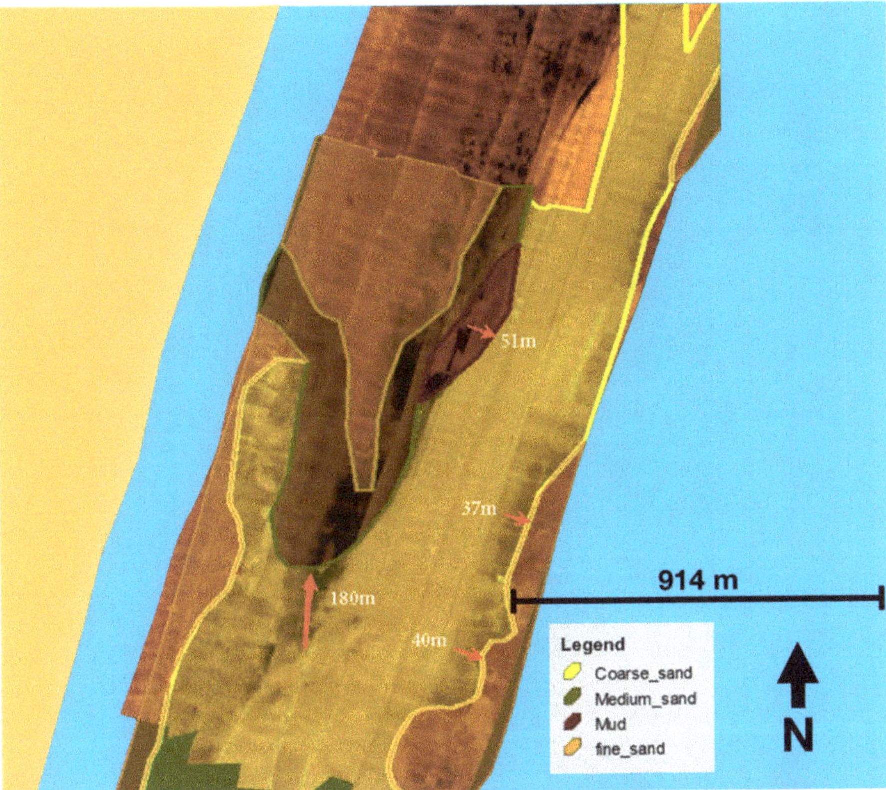

Figure 17. Temporal changes in sediment class boundaries, Assateague Island National Seashore, Maryland. The profile in the bottom is the side-scan sonar mosaics collected during May 2015. Shaded polygons on the top represent sediment classes inferred from the side-scan sonar mosaics collected during October 2015. Please note the northward shift (180 m) in the boundary of mud-coarse sand sediment class and Southeastward shift (37–40m) in the boundary of fine–coarse sand sediment class.

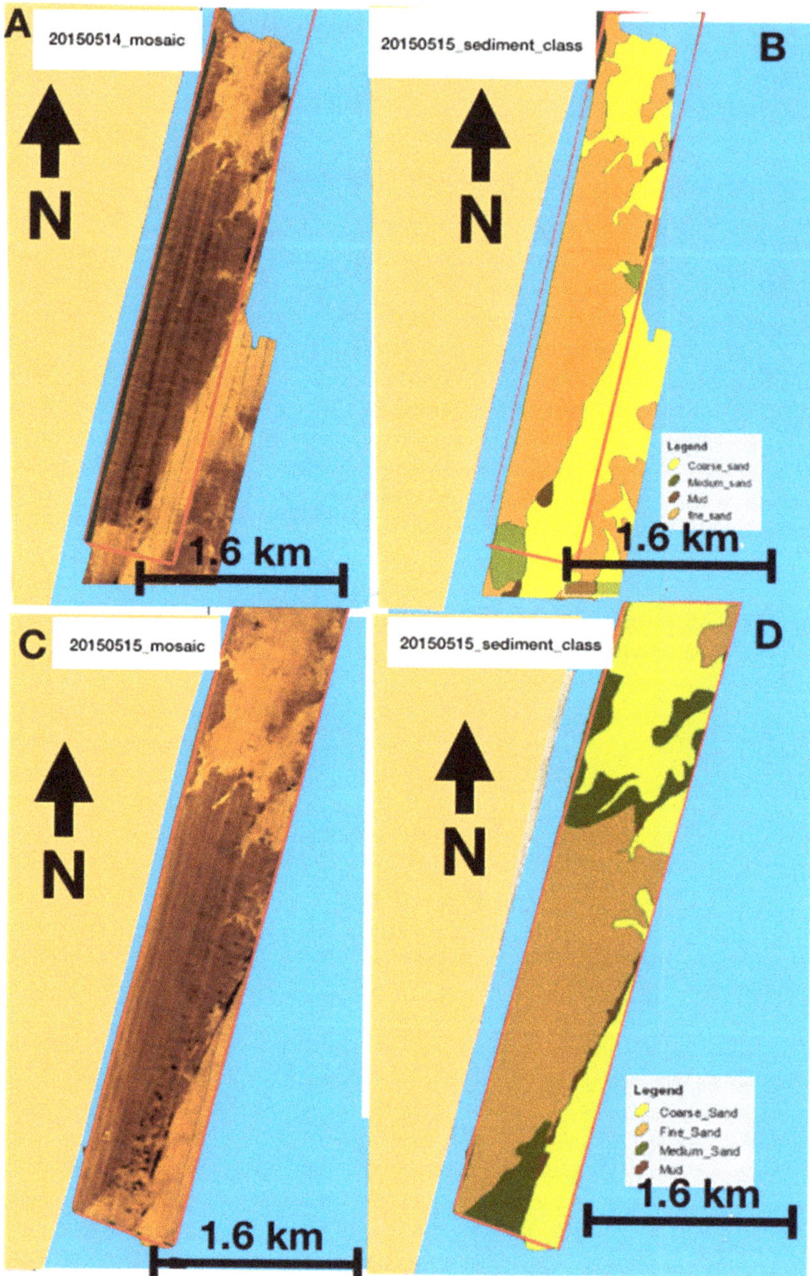

Figure 18. Side-scan sonar mosaic interpretation, Assateague Island National Seashore, Maryland. (**A**) Side-scan sonar mosaics collected during 14 May 2015; (**B**) interpretation of side-scan sonar data collected in 14 May 2015; (**C**) side-scan sonar mosaics collected in October 2015; (**D**) interpretation of side-scan sonar mosaics collected in October 2015. Note that red polygons represent the same areas mapped during both 14 May 2015 and October 2015.

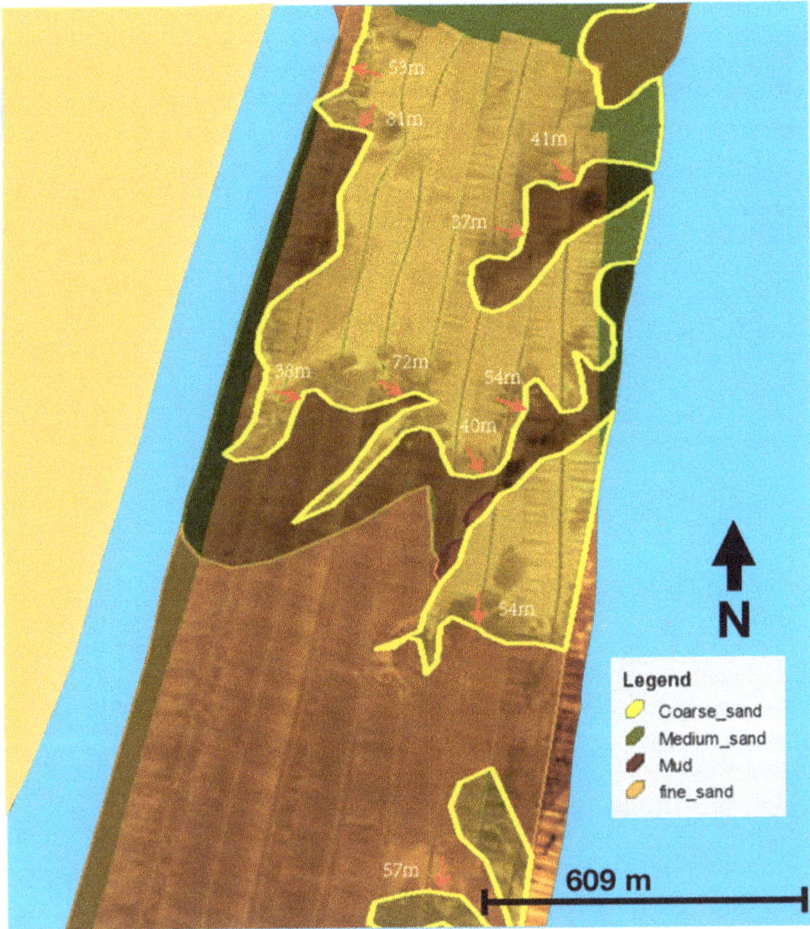

Figure 19. Temporal changes in sediment class boundaries, Assateague Island National Seashore, Maryland. The profile in the bottom is the side-scan sonar mosaics collected during May 2015. Shaded polygons on the top represent sediment classes inferred from the side-scan sonar mosaics collected during October 2015. Please note the west-south-eastward shift (38–57 m) in the boundaries of fine–coarse sand sediment classes.

3.3. Temporal Changes in Bathymetry

Storms and extreme weathering events [21,22] or human induced disturbances such as dredging [23] have the potential to change the nearshore bathymetry. These changes can have serious impacts to the marine flora and fauna that most marine environment protection efforts are focused on.

The purpose of the bathymetric change comparison is to see if it is possible to determine significant changes in the bathymetry after the passage of hurricane Joaquin, which passed through the area in late September and early October of 2015. A total of 18 bathymetric cross sections were constructed using Interpolate Line tool in ArcScene across areas with overlapping survey coverage from before (May 2015) and after (October 2015) the passage of hurricane Joaquin. Nine of the cross sections (A1, B1 to J1) were made from May 2015 bathymetry, and the other nine cross section (A2, B2 to J2) were constructed from October 2015 bathymetry data along the same trackline as their pairs in May 2015.

These bathymetry profiles were then exported as Excel spreadsheets. Sediment class type was added to each of the points along the bathymetry profiles based on the attributes from the acoustic sediment class maps discussed previously. Next, the Excel spreadsheets of the bathymetry profiles were imported into MATLAB R2017 (Natick, USA) to generate comparison charts between pairs of bathymetry profiles. Sediment types were coded from 1 to 4: 1 as mud, 2 as fine sand, 3 as medium sand and 4 as coarse sand. Bathymetry and sediment class comparison charts were generated by subtracting sediment type values of May 2015 profiles from October 2015 profiles, in order to understand if the seafloor sediments are coarsening or fining with the lateral shift in bathymetry (Figures 20 and 21).

In the areas where cross sections were constructed more or less parallel to the direction of sediment class boundary shift, southwestward lateral shift of sand bars (up to 60 m) were recognized in bathymetry cross section profiles (Figures 22 and 23). In the areas where cross sections were made more or less normal to the direction of sediment class boundary shift, vertical shifts of up to 65 cm are recognized (Figure 24). The largest vertical change was recognized in October 2015 profile J2 (Figure 24), which suggests axial incision of about 65 cm. This is the closest profile to the shoreline. Other profiles parallel to profile J2 suggests that vertical change in bathymetry seems to decrease with increasing distance from the shoreline.

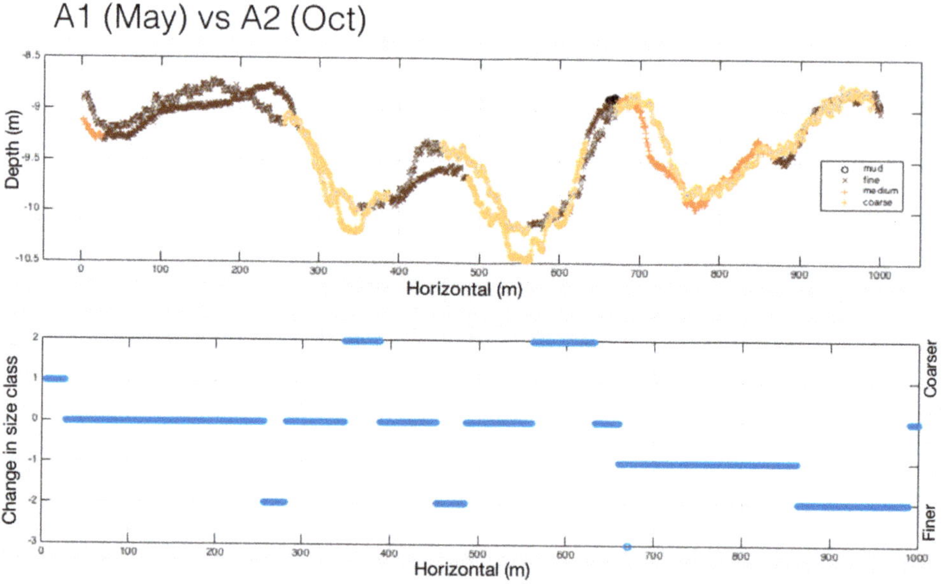

Figure 20. Bathymetry profile change versus sediment class boundary shift after Hurricane Joaquin. Curved bathymetry profiles A1 (May 2015) and A2 (October 2015) are color coded according to the sediment types in the chart located in the top. Horizontal scale are meter in both panels, and vertical are meters in the top panel and size class index in the bottom panel. Sediment classes were indexed to 4, 1 representing the finest and 4 representing the coarsest. The chart in the bottom was generated by subtracting sediment type values of A1 (May 2015) from A2 (October 2015) to understand if the seafloor sediments are coarsening or fining with the lateral shift in bathymetry. Please note the coarsening of sediments in the top of the sand bars and fining of sediments in the trough. Inset scale is in meters on both axes.

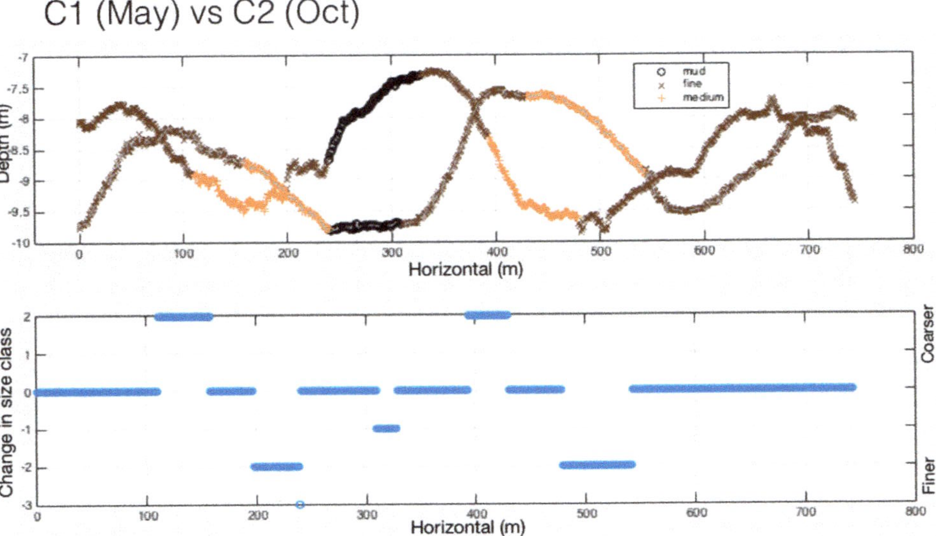

Figure 21. Bathymetry profile change versus sediment class boundary shift after Hurricane Joaquin, Assateague Island National Seashore. Curved bathymetry profiles C1 (May 2015) and C2 (October 2015) in Figure 25 are color coded according to the sediment types in the chart located in the top. Horizontal scale are meters in both panels, and vertical are meters in the top panel and size class index in the bottom panel. Sediment classes were indexed from 1 to 4, 1 representing the finest and 4 representing the coarsest. The chart in the bottom was generated by subtracting sediment type values of C1 (May 2015) from C2 (October 2015) to understand if the seafloor sediments are coarsening or fining with the lateral shift in bathymetry. Please note the coarsening of sediments in the top of the sand bars and fining of sediments in the trough. Inset scale is in meters on both axes.

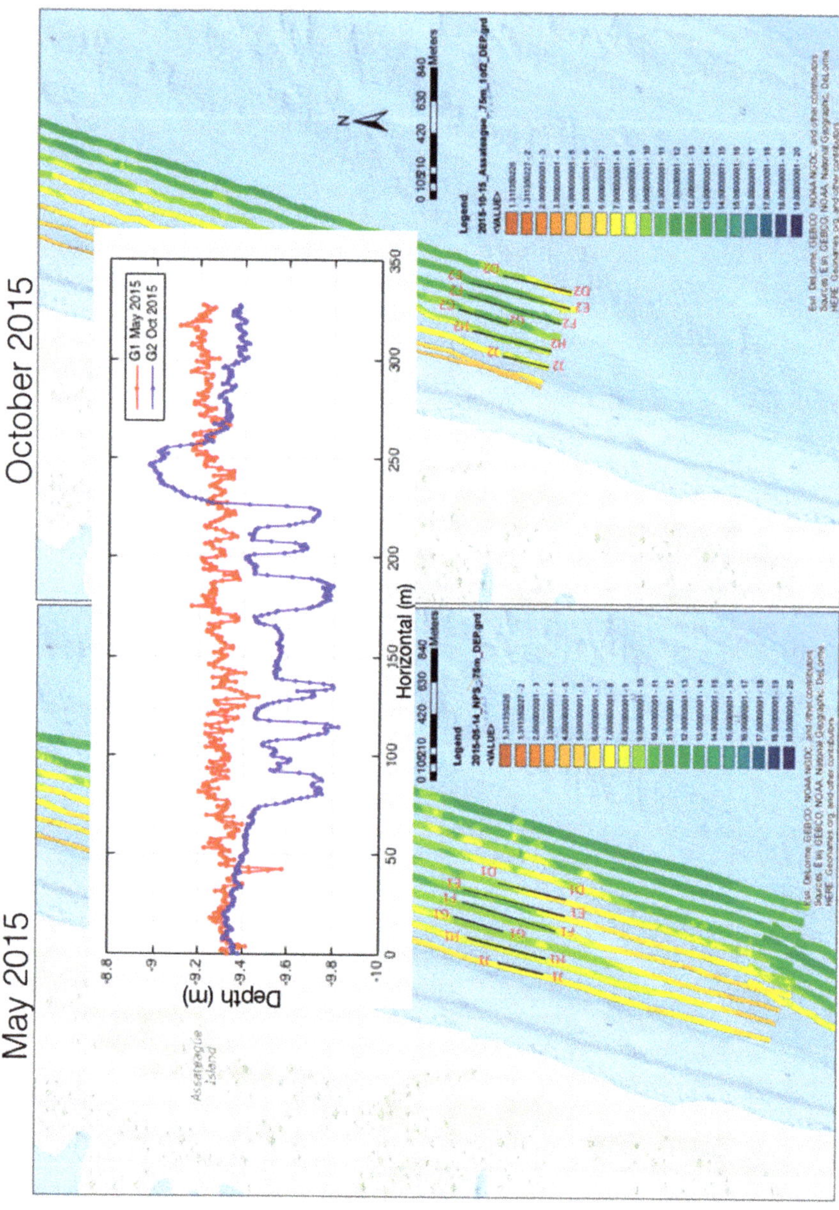

Figure 22. Bathymetry profile change after Hurricane Joaquin, Assateague Island National Seashore, Maryland. Red curved line represents the bathymetry profile C1 in May 2015 and blue curved line represents the bathymetry profile C2 in October 2015. Please note the Southwestward lateral shift of sand bars in the bathymetry profile. Inset scale is in meters on both axes.

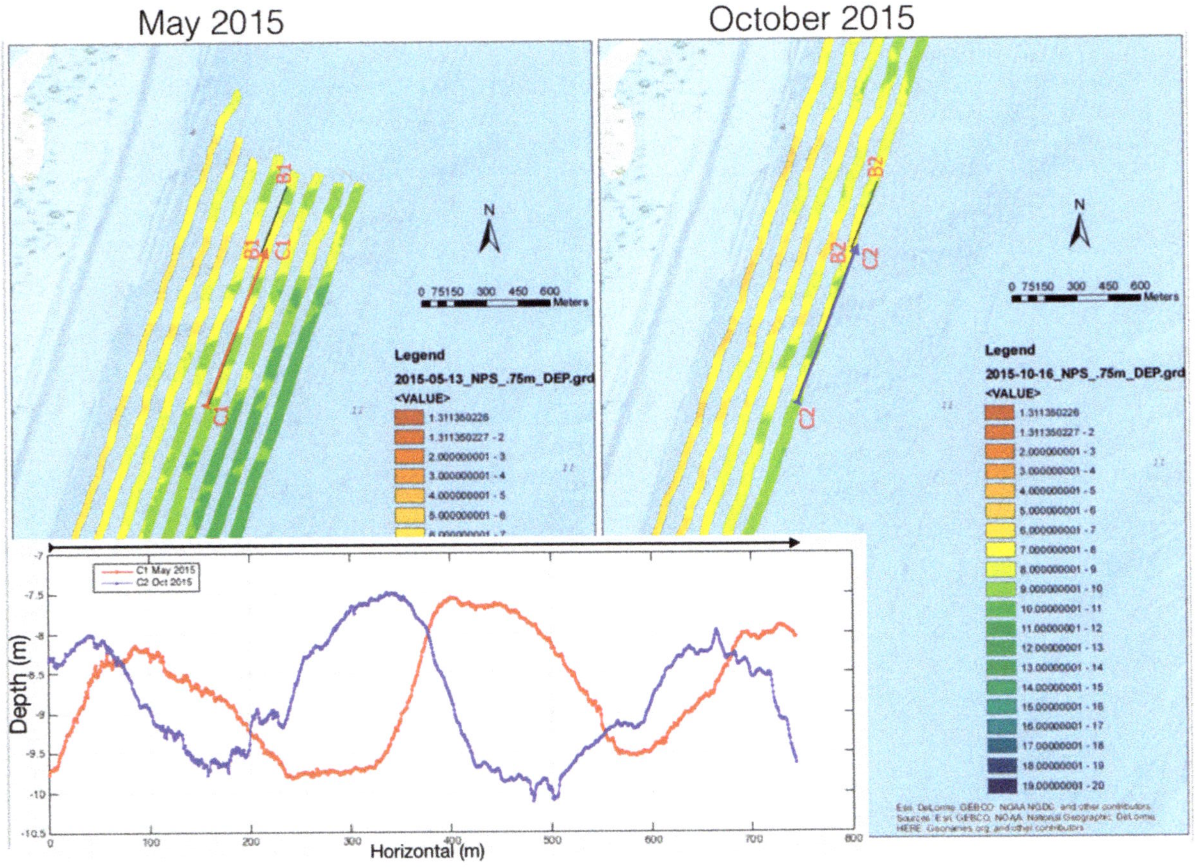

Figure 23. Bathymetry profile change after Hurricane Joaquin. Red curved line represents the bathymetry profile G1 in May 2015 and blue curved line represents the bathymetry profile G2 in October 2015. Please note the change in surface roughness and the vertical offset in the bathymetry.

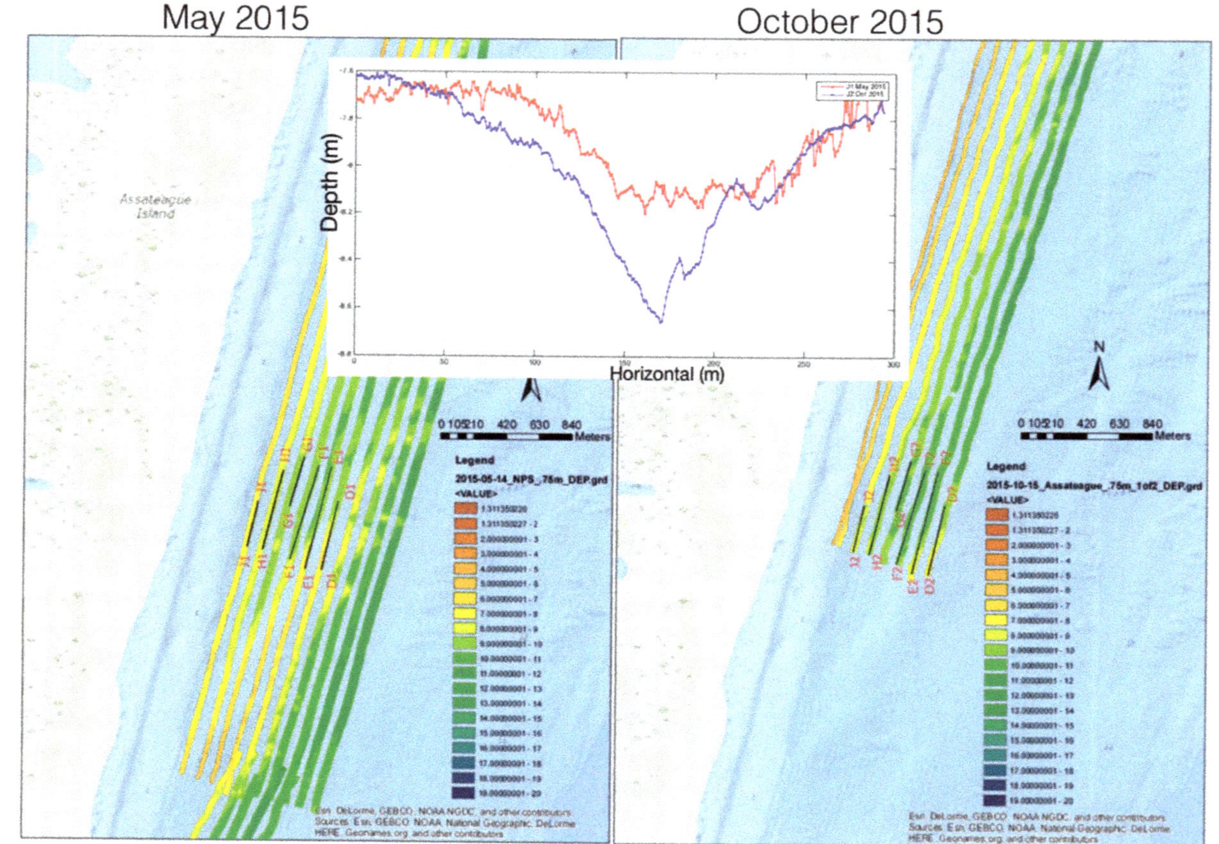

Figure 24. Bathymetry profile change after Hurricane Joaquin. Red curved line represents the bathymetry profile J1 in May 2015 and blue curved line represents the bathymetry profile J2 in October 2015. Please note the axial incision in J2 profile.

Apart from the axial incision recognized in October 2015 profile J2 (Figure 24), comparison of May and October profiles suggests that seabed changed from fairly smooth in May 2015 to incised by shore oblique/perpendicular bars in October 2015 (Figures 22 and 23). These are clear signs of hurricane effect on the seafloor. Decrease in vertical offset towards the sea may suggest an inverse relationship between hurricane effect on the seafloor and the distance from the shoreline. A Comparison between bathymetry change and sediment class boundary shift in this area suggests that sediments in the top of sand bars went coarser after the hurricane, while the ones in the trough of sand bars went finer (Figures 20 and 21).

4. Discussion

This nearshore benthic morphodynamic study at Assateague Island addresses a large gap in knowledge about the spatiotemporal dynamics of subtidal habitats and geomorphology in response to storm events (Figure 25). This knowledge will inform subsequent general management plans to best safeguard subtidal resources from anthropogenic and climate change stressors. This study also provides relevant data and recommendations for future assessments at ASIS.

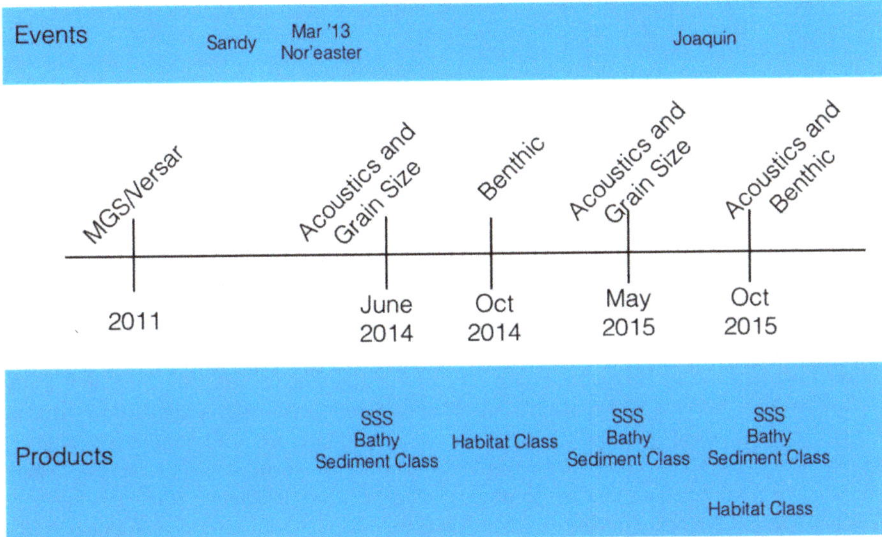

Figure 25. Timeline of Storm events and seafloor mapping studies and resulting benthic habitat and morphology products.

Although storms have varied morphologic impacts on different benthic substrates (i.e. coarse sand versus mud), the ASIS nearshore zone experienced large lateral shifts on the order of 30–70 m of sediment type contacts resulting in changes from fine to coarse sediment type with additional vertical depth changes from erosion and migration of between 50 cm to as much as 2 m associated with migration of nearshore oblique sand bars as a result the passage of storms.

We are not able to directly link the geomorphology and habitat changes outlined here to Hurricane Sandy, excluding the possibility of impacts from other storms throughout the overall study period. This is largely due to the study timeline (Figure 26) and constraints of sampling logistics around unexpected storm events. It is more realistic to characterize the sum of geomorphology and habitat shifts between the two surveys as integrated changes over a multi-year scale. Despite the inherent uncertainty in the study, this benthic morphodynamic study is a valuable resource to inform management of publicly-valued and historically under-studied habitats.

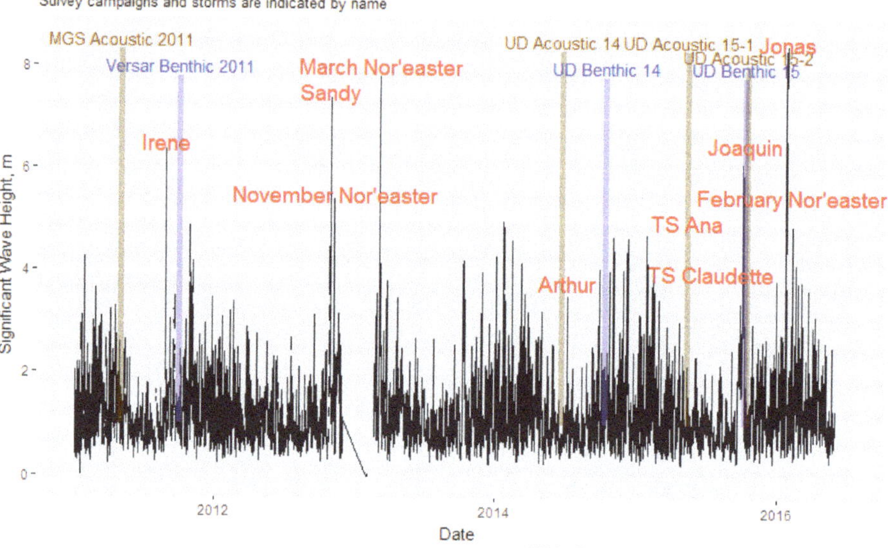

Figure 26. Significant wave height (H_s) in meters (m) at offshore NDBC Buoy 44009 (38.461 N 74.703 W) between 2011–2016 including the passage of hurricanes Sandy and Joaquin near the study site.

Spanning four years and a total of 22 storms—including four hurricanes between the two overall site surveys is the prime challenge facing our interpretation of these results and direct comparison of pre- and post-Sandy (2014–2015) benthic assemblage data. Typically, benthic storm-response studies must be conducted fairly soon before and after an event, in order to minimize any effects from additional weather events [6]. However, in this project we were unable to begin benthic mapping until almost two years after Hurricane Sandy made landfall in the region. While it is in the interest of the National Park Service who manages Assateague island to gain any and all knowledge of the impacts of hurricane Sandy on subtidal habitats and resources along ASIS, it is more reasonable to look at the post-Sandy (2014–2015) data as the results of continuous storm action, of varying degrees of magnitude, over the entire study period from October, 2011 through October, 2015.

The passage and impact of hurricane Joaquin in September 2015 afforded us a unique opportunity to conduct targeted post-storm surveys in areas that we had mapped in May 2015. This provided the basis for examining storm induced changes. The resulting shifts illustrated in the sediment type contacts in the side-scan sonar (Figures 11 and 12) and in the seabed morphology revealed in the bathymetry (Figures 21 and 23) indicate significant shifts in the seabed ranging between 20–180 m horizontally with bathymetric changes of as much as 2 m with the passage of shore oblique bars. In general, the shifts were in a South to Southwest direction consistent with the impacts from strong northeasterly winds and waves that accompanied the storm.

In terms of recommendation for future research efforts we would recommend the development and implementation of a targeted rapid response mapping program. The development of a short-term program for immediate storm-response surveys during the hurricane season at Assateague—while likely constrained to smaller spatial scales—could provide more accurate links between storms and benthic morphodynamics. Such a program would necessarily work on a reduced scale, targeting a select number of sites that are determined to undergo the most change from storm impacts.

Author Contributions: The authors contributions were as follows. A.T. was responsible for overall project conceptualization, resources, writing of the draft and revisions along with project administration and funding acquisition. A.A. was responsible for GIS analysis on bathymetric and sediment texture change, assisted in writing the draft and preparing figures. K.H. was involved in field data collection, data analysis and data curation. C.D. assisted in data acquisition, data analysis and provided input and guidance to the draft and revisions.

Funding: This study was funded by the National Park Service under CESU Task P14AC00380 and modification 14-0001.

Acknowledgments: Pre-storm data was provided by Versar, Inc. We thank the Assateague Island National Seashore staff and project partners for their input, including Monique Lafrance-Bartley, Bill Hulslander, Charles Roman, Mark Finkbeiner, Sara Stevens and Robin Baranowski. Field and lab work was made possible by Captains Kevin Beam and Evan Falgowski, Ellie Rothermel, Hannah Rusch, Danielle Ferraro, Stephanie Dohner, Tim Pileguard, Trevor Metz, Anna Schutschkow, Annie Daw, Jason Button, Leah Morgan, Drew Friedrichs, Rachel Dalafave, Alec Halbruner, Meghan Owings, Rachel Oidtman, and Alex Itin. We thank the Environmental Protection Agency Region III's Ocean and Dredge Disposal Program Team, especially Renee Searfoss and Sherilyn Morgan Lau, for the loan of a Young grab sampler, as well as the University of Delaware's Chris Sommerfield and Adam Marsh for the loans of additional laboratory and field equipment.

Conflicts of Interest: The authors declare no conflict of interest. The funders had no role in the design of the study; in the collection, analyses, or interpretation of data; in the writing of the manuscript, or in the decision to publish the results.

References

1. Balthis, W.L.; Hyland, J.L.; Bearden, D.W. Ecosystem responses to extreme natural events: Impacts of three sequential hurricanes in fall 1999 on sediment quality and condition of benthic fauna in the Neuse River Estuary, North Carolina. *Environ. Monit. Assess.* **2011**, *119*, 367–389. [CrossRef] [PubMed]

2. Engle, V.D.; Hyland, J.L.; Cooksey, C. Effects of Hurricane Katrina on benthic macroinvertebrate communities along the northern Gulf of Mexico coast. *Environ. Monit. Assess.* **2009**, *150*, 193–209. [CrossRef] [PubMed]

3. Douvere, F.; Ehler, C. Ecosystem-based Marine Spatial Management: An Evolving Paradigm for the Management of Coastal and Marine Places. *Ocean Yearb. Online* **2009**, *23*, 1–26. [CrossRef]

4. Goff, J.A.; Austin, J.A., Jr.; Flood, R.D.; Christensen, B.; Browne, C.M.; Saustrup, S. Rapid Response Survey Gauges Sandy's Impact on Seafloor. *Eos Trans. Am. Geophys. Union* **2013**, *94*, 337–338. [CrossRef]

5. Hapke, C.J.; Stockdon, H.F.; Schwab, W.C.; Foley, M.K. Changing the Paradigm of Response to Coastal Storms. *Eos Trans. Am. Geophys. Union* **2013**, *94*, 189–190. [CrossRef]

6. Trembanis, A.C.; DuVal, C.; Beaudoin, J.; Schmidt, V.; Miller, D.; Mayer, L. A detailed seabed signature from Hurricane Sandy revealed in bedforms and scour. *Geochem. Geophys. Geosyst.* **2013**, *14*, 4334. [CrossRef]

7. Trembanis, A.C.; Forrest, A.L.; Miller, D.C.; Lim, D.S.S.; Gernhardt, M.L.; Todd, W.L. Multiplatform ocean exploration: Insights from the NEEMO space analog mission. *J. Mar. Tech. Soc.* **2012**, *46*, 7–19. [CrossRef]

8. Miller, D.C.; Raineault, N.A.; Trembanis, A. Mapping Patchy Hard-Bottom Habitats with Multiple Technologies from Ship, ROV and Autonomous Underwater Vehicles in Delaware Bay. In Proceedings of the AGU Ocean Sciences Meeting, Honolulu, HI, USA, 20–24 February 2010.

9. Raineault, N.A.; Trembanis, A.C.; Miller, D.C. Mapping Benthic Habitats in Delaware Bay and the Coastal Atlantic: Acoustic Techniques Provide Greater Coverage and High Resolution in Complex, Shallow-Water Environments. *Estuaries Coasts* **2012**, *35*, 682–699. [CrossRef]

10. Diaz, J.R.; Solan, M.; Valente, R.M. A review of approaches for classifying benthic habitats and evaluating habitat quality. *J. Environ. Manag.* **2004**, *73*, 165–181. [CrossRef] [PubMed]

11. Raineault, N.A.; Trembanis, A.C.; Miller, D.C.; Capone, V. Interannual changes in seafloor surficial geology at an artificial reef site on the inner continental shelf. *Cont. Shelf Res.* **2013**, *58*, 67–78. [CrossRef]

12. Parnum, I.; Siwabessy, J.; Gavrilov, A.; Parsons, M. A Comparison of Single Beam and Multibeam Sonar Systems in Seafloor Habitat Mapping. In Proceedings of the 3rd Int. Conf. and Exhibition of Underwater Acoustic Measurements: Technologies & Results, Nafplion, Greece, 21–26 June 2009.

13. Hauser, O. Tubeworm Nodule reefs of Delaware Bay: An Evaluation of Two Acoustic Survey Methods and Lab Experiments of Nodule Formation. Ph.D. Thesis, University of Delaware, Newark, DE, USA, 2002.

14. Wilson, B.D.; Madsen, J.A. Acoustical methods for bottom and sub-bottom imaging in estuaries: Benthic mapping project to identify and map the bottom habitat and sub-bottom sediments of Delaware Bay. *Sea Technol.* **2006**, *47*, 43–46.

15. Holland, K.T.; Keen, T.; Kaihatu, J.M. Understanding Coastal Dynamics in Heterogeneous Sedimentary Environments. In *Coastal Sediments '03*; World Scientific Clearwater Beach: Pinellas, FL, USA, 2003.

16. Trembanis, A.C.; Wright, L.D.; Friedrichs, C.T.; Green, M.O.; Hume, T. The effects of spatially complex inner shelf roughness on boundary layer turbulence and current and wave friction: Tairua embayment, New Zealand. *Cont. Shelf Res.* **2004**, *24*, 1549–1571. [CrossRef]

17. Green, M.O.; Vincent, C.E.; Trembanis, A.C. Suspension of coarse and fine sand on a wave-dominated shoreface, with implications for the development of rippled scour depressions. *Cont. Shelf Res.* **2004**, *24*, 317–335. [CrossRef]

18. Carruthers, T.; Beckert, K.; Dennison, B.; Thomas, J.; Saxby, T.; Williams, M.; Fisher, T.; Kumer, J.; Schupp, C.; Sturgis, B.; et al. Assateague Island National Seashore Natural Resource Condition Assessment: Maryland, Virginia. In *Natural Resource Report NPS/ASIS/NRR—2011/405*; National Park Service: Fort Collins, CO, USA, 2011.

19. Carton, G.; DuVal, C.; Trembanis, A.C. A Risk Framework for Munitions and Explosives of Concern in Support of U.S. Offshore Wind Energy Development. *Mar. Soc. Technol. J.* **2019**, *53*, 6–20. [CrossRef]

20. Finlayson, D.P.; Miller, I.M.; Warrick, J.A. Bathymetry and Acoustic Backscatter—Elwha River Delta, Washington: U.S. Geological Survey Open-File Report 2011–1226, 2011. Available online: https://pubs.usgs. gov/of/2011/1226 (accessed on 10 October 2019).

21. Morton, R.A. Factors controlling storm impacts on coastal barriers and beaches—a preliminary basis for near real-time forecasting. *J. Coast. Res.* **2002**, *18*, 486–501.

22. Morton, R.A.; Sallenger, A.H., Jr. Morphological impacts of extreme storms on sandy beaches and barriers. *J. Coast. Res.* **2003**, *19*, 560–573.

23. Cooper, K.; Froján, C.; Defew, E.; Curtis, M.; Fleddum, A.L.; Brooks, L.; Paterson, D. Assessment of ecosystem function following marine aggregate dredging. *J. Exp. Mar. Biol. Ecol.* **2008**, *366*, 82–91. [CrossRef]

 Journal of
Marine Science and Engineering

Article

Application of Sentinel-2 Multispectral Data for Habitat Mapping of Pacific Islands: Palau Republic (Micronesia, Pacific Ocean)

Francesco Immordino [1], Mattia Barsanti [2], Elena Candigliota [1], Silvia Cocito [2], Ivana Delbono [2] and Andrea Peirano [2,*]

1 ENEA, Division Models and Technologies for Risks Reduction, 40129 Bologna, Italy
2 ENEA, Division Protection and Enhancement of Natural Capital, 19100 La Spezia, Italy
* Correspondence: andrea.peirano@enea.it

Received: 31 July 2019; Accepted: 31 August 2019; Published: 12 September 2019

Abstract: Sustainable and ecosystem-based marine spatial planning is a priority of Pacific Island countries basing their economy on marine resources. The urgency of management coral reef systems and associated coastal environments, threatened by the effects of climate change, require a detailed habitat mapping of the present status and a future monitoring of changes over time. Here, we present a remote sensing study using free available Sentinel-2 imagery for mapping at large scale the most sensible and high value habitats (corals, seagrasses, mangroves) of Palau Republic (Micronesia, Pacific Ocean), carried out without any sea truth validation. Remote sensing 'supervised' and 'unsupervised' classification methods applied to 2017 Sentinel-2 imagery with 10 m resolution together with comparisons with free ancillary data on web platform and available scientific literature were used to map mangrove, coral, and seagrass communities in the Palau Archipelago. This paper addresses the challenge of multispectral benthic mapping estimation using commercial software for preprocessing steps (ERDAS ATCOR) and for benthic classification (ENVI) on the base of satellite image analysis. The accuracy of the methods was tested comparing results with reference NOAA (National Oceanic and Atmospheric Administration, Silver Spring, MD, USA) habitat maps achieved through Ikonos and Quickbird imagery interpretation and sea-truth validations. Results showed how the proposed approach allowed an overall good classification of marine habitats, namely a good concordance of mangroves cover around Palau Archipelago with previous literature and a good identification of coastal habitats in two sites (barrier reef and coastal reef) with an accuracy of 39.8–56.8%, suitable for survey and monitoring of most sensible habitats in tropical remote islands.

Keywords: Sentinel-2; Remote Sensing; habitat mapping; mangroves; coral reefs; climate change; vulnerable habitats

---

1. Introduction

Coral reefs, seagrasses, and mangroves are threatened worldwide by climate change, whose main effects are sea temperature increase and ocean acidification [1–3]. In addition, the frequency of discrete extreme warming events (heat waves) threatening global biodiversity has increased, with projections indicating they will become more frequent [4]. Moreover, coastal coral ecosystems are threatened by additional anthropogenic pressures as overfishing, urbanization, and tourism [5]. Consequences are producing concerns for the maintenance of ecosystem functioning and the associated flow of services that coral reefs provide.

Having a deep cultural heritage for ocean conservation, the Pacific Ocean countries are strong advocates of a 'Blue Economy Strategy' and the sustainable use of ocean resources for economic growth, and are turning towards an increased reliance on green tourism. Palau Republic, among others, has emerged as a global leader in ocean conservation, receiving the 2012 Future Policy Award for developing the world's best policies to protect oceans and coasts. People living in Pacific Islands depend on healthy coastal ecosystems for their survival. Ecosystems such as coral reefs, mangroves, and seagrass beds favor coastal protection, provide food, building materials, and they represent the principal economic incomes for fishing and tourism industries [5]. Hermatypic corals are the most sensitive organisms to the synergic effects of warming, hurricane destruction, and ocean acidification [1]. Future climate scenery predicts that over the coming decades coral mortality may reach up to 60% in the areas where shallow coral reefs are present [6], driving to the elimination of most warm water coral reefs by 2040–2050 [1]. Hence, the preservation of global reef biodiversity, the monitoring of climate change effects on reef and islands together with the development of global strategies to reduce greenhouse gas emissions are among the major management issues to counter the effects of climate change [7].

The need to effectively manage coral reef systems and their associated coastal environments, such as mangroves and seagrasses, requires the ability to document their present status and monitor changes over time. Benthic habitat mapping of coastal ecosystems is an important and essential mean to provide marine resource assessments for coastal management and ecological analysis. Habitat mapping by remote sensing allows large scale environmental patterns and is highly cost-effective compared to the sampling of physical areas achievable by field survey, which provide accurate data but at highly detailed scales [8–13]. RS imagery were used in conjunction with state-of-the-art RS algorithms to map reefs geomorphology and habitat distribution [14–17].

Today, widely available orthorectified satellite imagery (Google Earth) and rapid development and cost reduction in GIS, multibeam echosounders, Lidar, and RS technologies are making large-scale morphometric quantification of reefs feasible [18–20].

In 2001, the "Millennium Coral Reef Mapping Project—Understanding, Classifying and Mapping Coral Reef Structures Worldwide Using High Resolution Remote Sensing Spaceborne Images" examined ~1500 images to design a thematically rich (966 classes) geomorphological classification scheme, used to interpret and map every single reef of the planet. Distributed as Geographic Information System (GIS) layers in late 2003, the map products have been used for a variety of applications, from establishment of marine protected areas in Papua New Guinea and Eastern Caribbean, reef condition assessment in the Caribbean, morphometric analyses of Maldivian reefs, and geochemical budgets in French Polynesian atolls [21].

Sentinel mission is one of the last RS programs dedicated to the study of marine and coastal environment. A constellation of two satellites, Sentinel-2 A and B, was launched in 2015 by European Space Agency (ESA) as part of the Copernicus Program. The Sentinel-2 satellites carries an innovative wide swath high-resolution multi-spectral imagery with 13 bands (443–2190 nm) in the visible, near infrared, and short wave infrared part of the spectrum. The combination of high spatial resolution of 10–20–60 m, novel spectral capabilities, a swath width of 290 km, a global coverage of land surfaces from 56° S to 84° N and frequent revisit times (5–15 days) provides unprecedented views of Earth. Sentinel mission not only offers continuity of services for the moderate resolution multispectral Spot XS and Landsat Thematic Mapper series sensors, but it also has several technical improvements that may lead to enhanced capability in coral reef mapping applications [22–24].

The aim of this paper is to enhance the effectiveness of RS repositories as a powerful tool for coastal resources assessment and management of remote Pacific islands, where mapping data are missing or lacking at present. A low-cost methodological approach for a preliminary habitat classification useful for coral coastal management purposes is proposed for Palau Archipelago (Micronesia). Different available resources such as free satellite images from the Copernicus Open Access HUB services and ancillary information as Google Earth imagery, scientific reports and literature were used to create maps of relevant habitats without field data validation. Results on mangrove forest distribution around the islands and on habitat classes in two sites (barrier reef and coastal reef) are compared with available detailed habitat maps of the Palau coastal area, and pros and cons of this approach are discussed.

2. Materials and Methods

2.1. Study Area

The Republic of Palau is an island country located in the western Pacific Ocean; the country contains approximately 340 islands, forming the western chain of the Caroline Islands in Micronesia, and has an area of around 466 square kilometres (Figure 1).

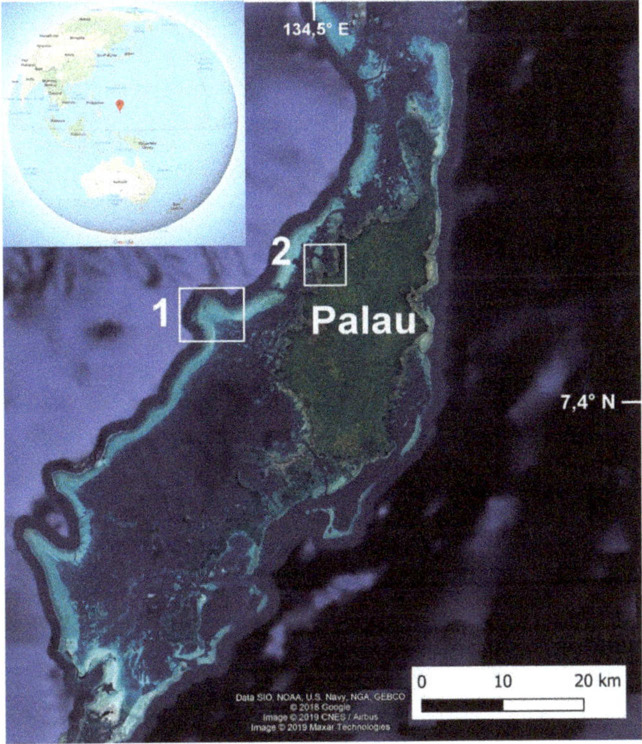

Figure 1. The main island of the Palau Republic (Micronesia—West Pacific). Numbers indicate the barrier reef zone (site no. 1) and coastal reef zone (site no. 2) where the shallow-water benthic habitat mapping was performed (basemap and inset Google © 2019).

The Palauan coral reef ecosystem has the most diverse flora and fauna of Micronesia. Palau has some of the most extensive seagrass beds in all of Micronesia, hosting 10 species of seagrass [25]. Seagrasses are valuable habitats that provide important ecological components of coastal ecosystems worldwide. Moreover, one of the most significant ecosystems in Palau are the mangrove forests, the transition zone between terrestrial and marine ecosystems. The most extensive areas of mangrove forests occur along the west coast of the main island of Palau Republic (Babeldaob), covering approximately 80 percent of the shoreline. In the mangrove forest of Palau, there are 18 mangrove trees and associated plant species, the most diverse in Micronesia. These habitats provide a number of ecological functions, from nurseries for juvenile fish to food and shelter for invertebrates and rare, protected species as sea turtles, crocodiles, and dugongs. Mangroves are ecologically important also because they help stabilize coastal areas by trapping and holding sediments washed down from inland areas and local watersheds. Moreover, the islands of the Pacific region lie in one of most seismically active regions of the world and for the coastal population mangrove forests are often the first line of defence against such natural calamities.

2.2. Image Processing and Shallow-Water Benthic Classification

Two Sentinel-2 2017 satellite images (4 July 2017) were downloaded from the open source portal of the European ESA-Copernicus project (https://scihub.copernicus.eu/). Sentinel-2 product used in this work provides orthorectified Top-Of-Atmosphere (TOA) reflectance (Level 1-C), with sub-pixel multispectral registration. Cloud and land/water masks are included in the product. SENTINEL-2 products are available to users in SENTINEL-SAFE (Standard Archive Format for Europe) format, including image data in JPEG2000 format, quality indicators, auxiliary data, and metadata (https://sentinel.esa.int/web/sentinel/user-guides/SENTINEL-2-msi/data-formats). The SAFE format has been designed to act as a common format for archiving and conveying data within ESA Earth Observation archiving facilities. The SAFE format wraps a folder containing image data in a binary data format and product metadata in XML. This flexibility allows the format to be scalable enough to represent all levels of SENTINEL products. The SENTINEL-2 product refers to a directory folder that contains a collection of information and includes: manifest.safe file which holds the general product information in XML; preview image in JPEG2000 format; subfolders for measurement datasets including image data (granules/tiles) in GML-JPEG2000 format; subfolders for datastrip level information; subfolder with auxiliary data (e.g., International Earth Rotation & Reference Systems (IERS) bulletin); HTML previews.

The ERDAS ATCOR radiative transfer model was used for atmospheric corrections (Figure 2): it eliminates atmospheric and topographic effects in satellite imagery and extracts physical surface properties, such as surface reflectance, emissivity, and temperature. ATCOR Workflow employs a database containing the result of radiative transfer calculations based on MODTRAN® 5.

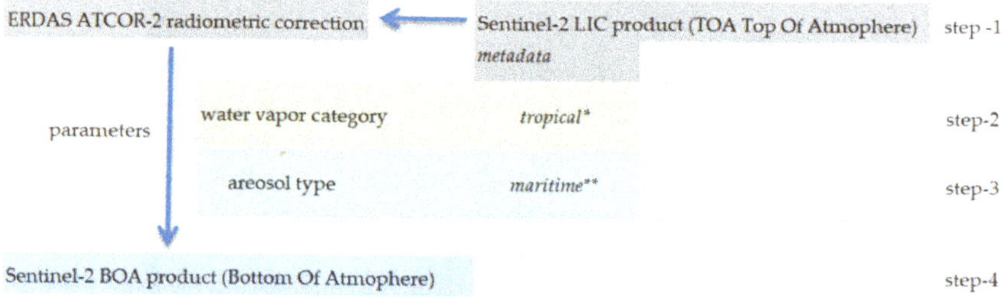

Figure 2. Atmospheric correction steps: ERDAS ATCOR radiative transfer model used for Sentinel-2 L1C product. * Tropical corresponds to a water vapor column of 4.11 cm at sea level, while ** Maritime represents the aerosol conditions in areas close to the sea.

The processed multispectral satellite images showed the presence of calm, clear waters, and—considering the shallow depths—water penetrating correction and sun glint corrections procedures were not applied [26,27]. To this end, standard atmospheric corrections have been carried out, considering the type of atmospheric column, the SENTINEL-2 image acquisition period and the sensor instrument parameters that the ERDAS software automatically recognizes by reading the metadata at the beginning of the acquisition procedure.

All Sentinel-2 bands were processed and 're-sampled' in order to get the highest resolution possible (10 m); then the ENVI image analysis software was used for classification procedures.

Two separate approaches were used to classify inland mangrove cover and habitat classes related to coral platform; the analysis processing chain is illustrated in Figure 3.

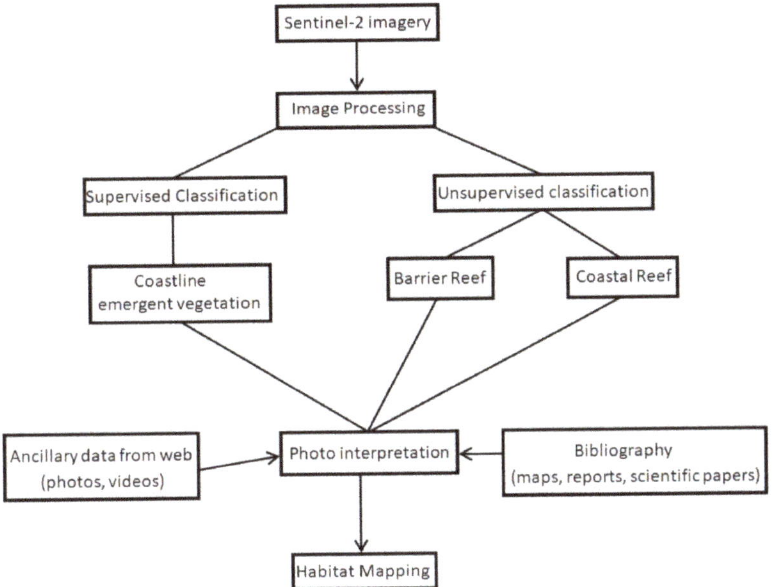

Figure 3. Workflow diagram illustrating the steps followed for the habitat mapping of Palau.

First, for the 'coastline emergent vegetation' or mangrove classification, all Sentinel-2 13 bands were processed and 'resampled', building a metafile, in order to get the highest resolution possible (10 m). Then Band 11 (wavelength range 1542–1685 nm) was particularly enhanced in order to distinguish the land, habitat of mangroves, and sea (Figure 4).

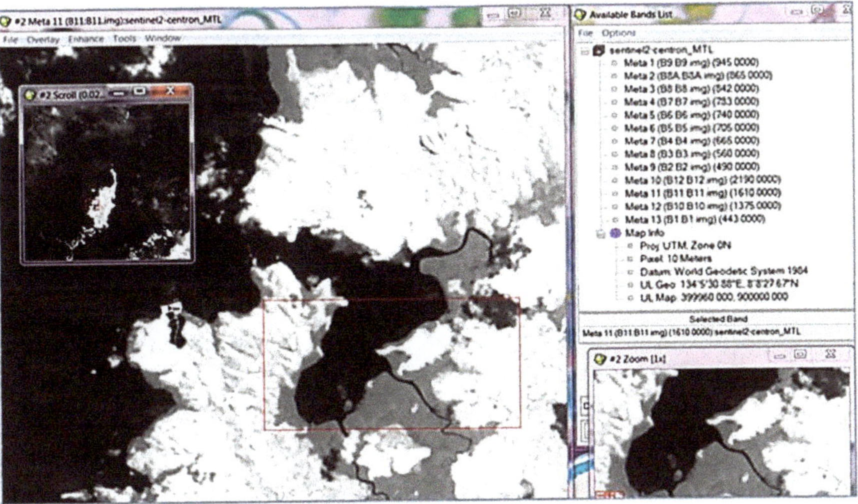

Figure 4. Sentinel-2 satellite image on the Republic of Palau on 4 July 2017, with a 10 m spatial resolution. Enhanced Band 11 was used, showing the land (white), the mangrove cover (grey), the sea (black).

The spectral signatures of mangroves and land vegetation were extracted from the Sentinel-2 image and here reported to show their spectral separability (Figure 5).

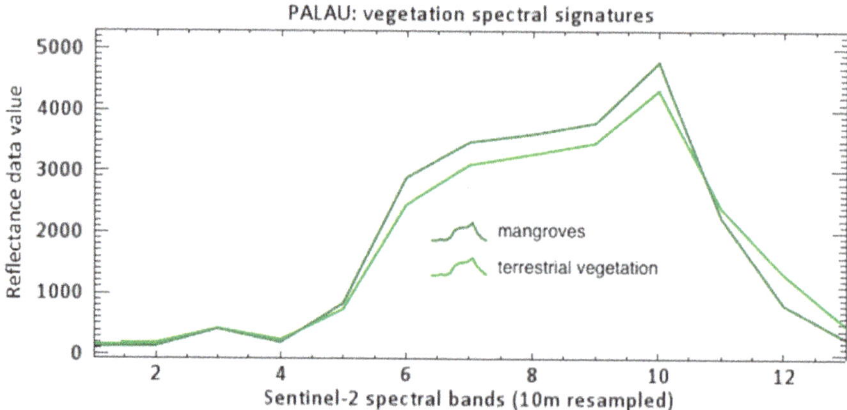

Figure 5. Spectral signatures of mangroves and terrestrial vegetation in Sentinel-2 spectral bands.

Hence, the ENVI 'supervised classification' through 'region of interest' (ROI) and 'maximum likelihood' methods were applied in order to separate the land from mangroves and sea. This approach allowed to compute the total shoreline length and the mangroves cover for the whole Palau Archipelago (Figure 6).

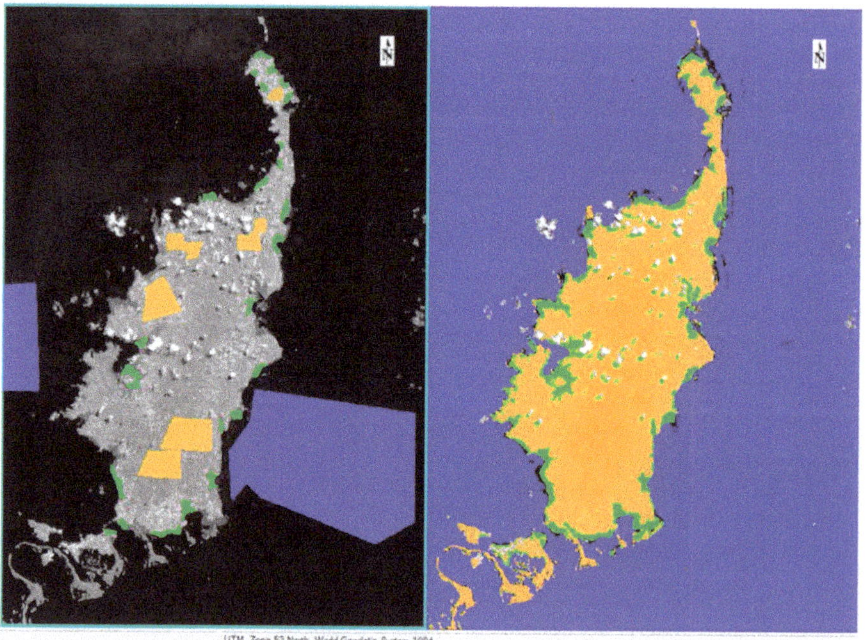

Figure 6. Sentinel-2 satellite image (Band 11) on the Republic of Palau on 4 July 2017, with a 10 m spatial resolution: the ROIs in the process of a 'supervised classification' (on the left); the classification result (on the right) where land is orange, the mangrove cover is green, the sea is blue. Clouds are isolated and labelled as unclassified.

For the habitat classification related to coral platforms, we restricted the analyses to two sites, located in the western side of the main Palau island (Figure 1) and representatives of the two main morphological zones of the island: the barrier reef (site no. 1) and the coastal reef (site no. 2). The ENVI algorithm that showed better results was the 'unsupervised isodata classification' performed for a maximum of 20 classes on Sentinel-2 using bands 2-3-4-8 with 10 m of resolution. Unsupervised classification yields an output image in which a number of classes are identified and each pixel is assigned to a class [28]. This classification often results in too many land cover classes, particularly for heterogeneous land cover types, and classes often need to be combined to create a meaningful map; the classification is useful when there is no pre-existing field data or detailed aerial photographs for the image area, and the user cannot accurately specify training areas of known cover type. In marine habitat classification, the unsupervised isodata classification is considered the most appropriate for an approach with no sea truth validation [29–31].

The spectral signatures of the reef coverage were extracted from Sentinel-2 image and here reported on a plot (Figure 7) to show the spectral separability among them. To check the reliability of atmospheric corrections and subsequent classification of the Sentinel-2 image, the ROI of the coverage classes present in the coralligenous environment was extracted.

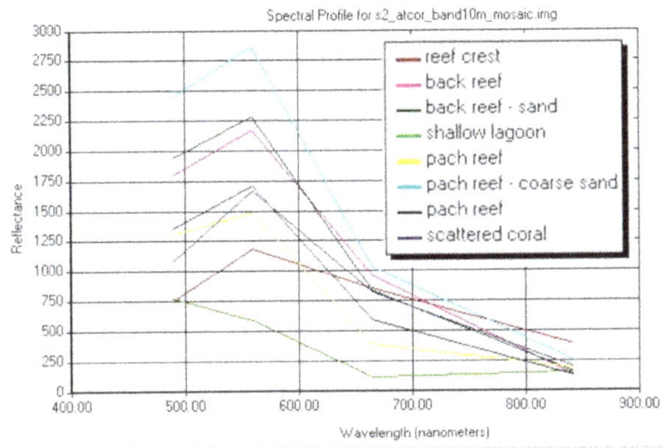

Figure 7. Spectral signatures of reef coverage (the barrier reef zone, site no. 1) in ERDAS software.

The image interpretation for the coastal benthic classification was based on two main components: the geomorphologic description of the seabed and the biological relevance of the associations living in the zone (Table 1).

We referred to the NOAA (National Oceanic and Atmospheric Administration, Silver Spring, MD, USA) Shallow-Water Benthic Habitats Manual [32] to identify priority habitats in terms of ecological function and coral presence. Hence, on the barrier reef, the following five main habitat classes were used for RS classification:

(1) The 'fore reef' is the outward part of the reef barrier. On its underwater cliff, all coral biodiversity is concentrated along the first 20–40 m of depth. It represents the most important reservoir for coral maintenance and survival; (2) The 'reef crest' is the part of the outer barrier reef more exposed to open-ocean waves. Its associated subclasses are the living coral and one algal ridge formed mainly by coralline algae. It is the most important area of the barrier for the defence of the coastline and it is the most sensitive to mortality due to low tides, elevated seawater temperatures, and storms; (3) The 'back reef' is the area of the barrier reef formed by a coral platform and is limited towards the coastline by the lagoon. It may be very large and in its shallow areas is characterized by an eroded platform and rubbles with associated subclasses of coralline algae, massive corals, and algal turf. It is the area where fragments of corals of the reef crest damaged by open-ocean waves may survive. The slope of the back reef limited by the lagoon is normally formed by sand and coral rubbles; (4) The 'patch reef' includes as subclasses the scattered living coral formations like coral knob, coral head, irregular little islands of aggregated corals surrounded by sand, found on the back reef, on coral platform, or dispersed in the

lagoon. The patch reef is important because it includes isolated living reef that represent a 'reservoir' both in term of larvae and individuals of different coral species; (5) The 'lagoon' is the area between the back reef and the coastline. Here coral, sand, and rubble are abundant. However, the lagoon represents a limit for the RS investigations when the depth may exceed 30 m. In this deep area, indicated as 'deep lagoon', the coral cover may be important but should be investigated with other methodologies (transects, multibeam, side scan sonar). The species of coral inhabiting the lagoon and the deep lagoon could be an important reservoir of larvae for the surrounding communities.

Table 1. Geomorphological description and biological importance of the main classes/subclasses used for the Palau habitat mapping with Sentinel-2 imagery.

Zone	Habitat Classes	Habitat Subclasses	Geomorphological Description	Biological Relevance
Barrier Reef	Bank/shelf and fore reef	coral, coralline algae	underwater coral cliff	dense coral, high biodiversity, coral reservoir
	Reef crest	coral, coralline algae	windward coral platform shelf edge, algae ridge	dense coral, high biodiversity
	Back reef	scattered coral, flesh algae	coral platform, sand channels	medium to scarse coral density
	Patch reef	coral/coralline algae	coral knob, aggregate coral	medium coral density, coral reservoir
	Lagoon	sand coral knoll, massive coral	sand, rubbles, coral	medium to dense coral density, coral reservoir
Coastal Reef	Reef crest	coral, coralline algae	seaward coastal coral platform, shelf edge	high coral density, high biodiversity, coral reservoir
	Reef flat	coral, flesh algae, seagrasses	coral platform	high coral density, high biodiversity, coral reservoir
	Uncolonized	outward limit of mangrove area	sand/mud	nursery area for fish, shrimps, etc.
	Emergent vegetation	mangrove	intertidal/sand/mud	high primary productivity area

On the coastal reef, the following three main habitat classes were used:

(1) The 'coastal fringing reef' or 'coastal reef crest' is the seaward fringe of the coastal reef flat. Its associated subclasses are living corals in good health; this habitat class is subjected to erosion by waves and it is the most important indicator of the coastal reef erosion together with blue holes; (2) The 'coastal reef flat' includes the reef platform formed by dead coral surrounding the coast of the major island and it can reach the coastline as a rock substrate. Its subclasses include seagrasses and algae (mainly macroalgae) in the area where fine sediments are deposited. Its importance is related to the coastal defence from erosion; (3) Shoreline and emergent vegetation is formed from the emergent vegetation habitat composed primarily of red mangroves, generally found in areas which are sheltered from high-energy waves.

Finally, the classes 'unknown' or 'unclassified' represent the uninterpretable areas due to cloud shadow, water depth, or other interference.

2.3. Validation of Image Interpretation

To validate the Sentinel-2 image interpretations, the habitat mapping results were compared with available Palau NOAA maps (https://products.coastalscience.noaa.gov/collections/benthic/e102palau/). These maps are the final results of a 4-year project [33]: NOAA collected 2002 and 2004 multispectral Ikonos and supplemental Quickbird satellite imagery of Palau Archipelago with spatial resolution of 4 m (raw multispectral) and 1 m (pan sharpened). Color balanced imagery proved suitable for visual extraction of the habitat classes. NOAA image process included atmosphere correction, deglinting, color balancing, orthorectification, correction for water column effects and pan sharpening. Collection constraints were set to control environmental effects such as glare, glint and other interferences that would limit visualization of benthic features. NOAA multiple collects were conducted to mosaic multiple scenes up to a maximum of 10% cloud cover. These images were used by NOAA to manually interpret and delineate geomorphologic features, cover types, and habitat boundaries. NOAA maps were produced following the schemes for habitat classification prepared by coral reef biologists and mapping experts. Ground validation information was used to investigate uncertainties on the photo-interpreter behalf during the decision-making process of the manual delineation of zones or structures. To test the interpretation accuracy, sea-truth validation was performed by NOAA through 623 benthic habitat characterizations conducted in four areas of Palau. Results showed an overall accuracy in habitat identification from 88.4 to 97.3% and with a thematic accuracy of the habitat classification schemes ranging from 0.79 Tau for detailed cover (79.9% overall accuracy) to 0.96 Tau for major structure (97.3% overall accuracy).

In the present work, main habitat classes cover derived from Sentinel-2 data were compared with NOAA and used as reference data for accuracy evaluation of the proposed method.

3. Results

The shoreline length and the mangrove cover were calculated for the whole Palau Archipelago. A comparison between NOAA results [28] on 2003–2006 Ikonos imagery and the present work on 2017 Sentinel-2 images is shown in Table 2. Shoreline and mangroves area values show relevant differences in the considered period, equal to 18.8% and 25.0%, respectively. The shoreline length of the Republic of Palau is calculated by NOAA, derived from 2003–2006 Ikonos Imagery, as 1021 km by visual interpretation and manual delineation of satellite imagery.

Table 2. Length of shoreline (km) and mangrove areas (ha) on Palau Archipelago computed in the present work (ENEA) and by NOAA (Analytical Laboratories of Hawaii, 2007). Differences are in percentages (%).

Zone	ENEA	NOAA	ENEA vs. NOAA
	2017	**(2003–2006)**	**(%)**
Shoreline (km)	1213	1021	+18.8
Mangroves area (ha)	5500	4400	+25.0

Regarding the mangrove habitat, thanks to the Band 11 of the Sentinel-2 imagery, the mangrove cover for the whole Palau Archipelago was determined with a resolution of 10 m and with a great accuracy. The optimum classification is mathematically confirmed by the Jeffries-Matusita distances among classes: the output is shown in Table 3. Good coefficients range between 1.999 and 2.000 [34].

Table 3. ROI matrix separability, for the SENTINEL-2 imagery, using Jeffries-Matusita algorithm.

Zone	Jeffries-Matusita Coefficients
Land–Sea	2.000–2.000
Sea–Mangroves	2.000–2.000
Land–Mangroves	1.999–1.999

As well as mangroves, also for coastal benthic habitats, Jeffries-Matusita (JM) spectral separability coefficients (Table S1, Supplementary Materials) confirm values between 1.9–2.0 among the different classes, with the only exception for the combination back–reef sand–algae and patch reef coral–sand with a JM value of 1.8, probably linked to the sedimentological features and texture of these classes due to the presence of biodetritic sand in the two classes. The image processing steps confirm the classes spectral reflectance, since bottom-types are statistically separable and identifiable on the base of their reflectance spectra [35]. The scatter plot shows a high spectral separability among coral reef classes (Figure 8).

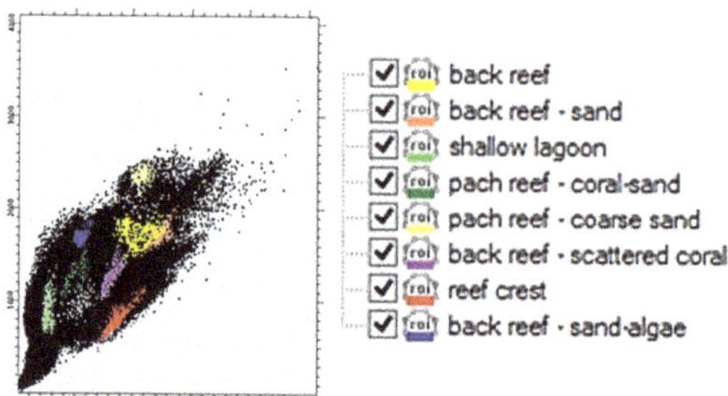

Figure 8. Scatter plot of the coral reef class coverage, showing a high classes separability.

Unsupervised isodata classification allowed the identification of 12 habitat classes/subclasses for the barrier reef (site no. 1) and 10 classes/subclasses for the coastal reef (site no. 2) (Figures 9 and 10). To test the goodness of the habitat classification performed on Sentinel-2 images, data were compared with NOAA habitat maps from Ikonos satellite imagery and sea-truth validations carried out in the period 2002–2004 [33]. Comparison of habitat classifications on the barrier reef (Figure 9) shows how the classification with Sentinel-2 image was not able to identify the class bank/shelf escarpment, i.e., the deeper area of the outward barrier found on NOAA images. In Sentinel-2 classification, the whole area was identified as fore reef. The reef crest and the back reef showed differences between NOAA findings [33] and our results. The area of reef crest was thinner and better defined in NOAA maps, wider in our estimation. On the contrary, the back reef and the patch reef areas were more defined in Sentinel-2 imagery where both scattered corals and patch reef were identified as well as the nature of the bottom (sandy or hard bottom).

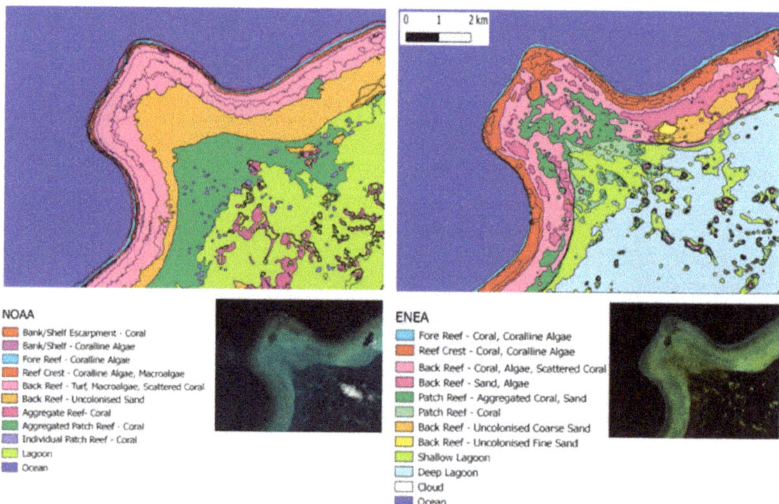

Figure 9. Comparison between habitat classifications in the barrier reef zone (site no. 1, see Figure 1). On the left, NOAA's map and 2004 Ikonos imagery (informative layers downloaded from https://products.coastalscience.noaa.gov/collections/benthic/e102palau/); on the right, present work (ENEA) habitat map and 2017 Sentinel-2 imagery.

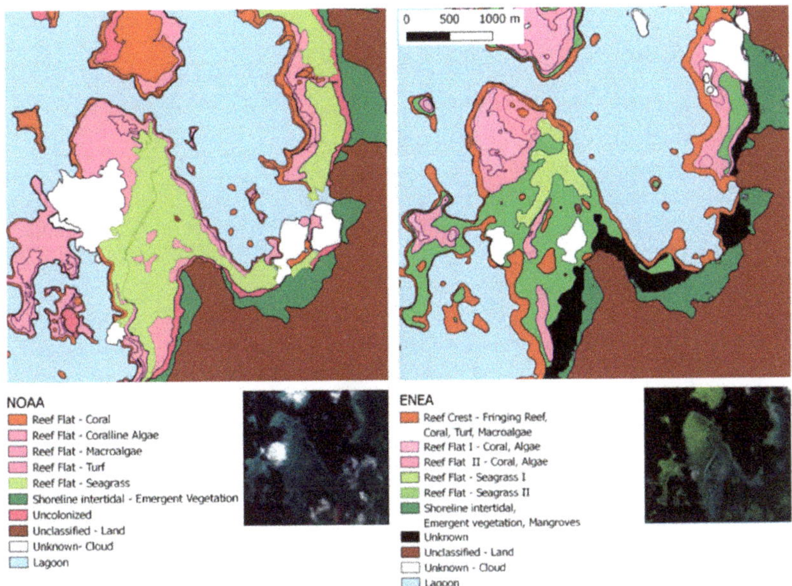

Figure 10. Comparison between habitat classifications in the Coastal Reef zone (site no. 2, see Figure 1). On the left, NOAA's map and 2004 Ikonos imagery (informative layers downloaded from https://products.coastalscience.noaa.gov/collections/benthic/e102palau/); on the right, present work (ENEA) habitat map and 2017 Sentinel-2 imagery.

The extension in hectares (ha) of the three main barrier reef areas (Table 4) showed an overestimation of Sentinel-2 classification of 12.3% for bank/shelf–fore reef and 20.0% for reef crest and back reef areas, differently an underestimation of lagoon area of 20.3%.

Table 4. Palau: Habitat classes cover (ha) in the two zones of the barrier reef (site no. 1) and coastal reef (site no. 2) computed for the 2017 in the present work (ENEA) and by NOAA (Analytical Laboratories of Hawaii, 2007) in 2004. Differences in accuracy are in percent (%).

Zone	Habitat Classes	ENEA (ha)	NOAA (ha)	ENEA vs. NOAA Overall Accuracy (%)
	Bank/shelf and fore reef	12,003	10,688	52.9
Barrier reef	Reef crest and back reef	40,568	33,794	54.6
	Lagoon	9512	11,940	44.3
Coastal reef	Reef crest	1555	2351	39.8
	Reef flat	4231	3222	56.8

The comparison for the coastal reef area of Palau (Figure 10) showed the suitability of Sentinel-2 images for the recognition of the main habitat classes on the reef platform found by NOAA in 2004. Both the reef crest (or fringing reef) and reef flat were recognized in Sentinel-2 images and classification.

The extension in hectares (ha) of the three main coastal reef areas (Table 4) showed an underestimation of Sentinel-2 classification of 33.9% for the reef crest and an overestimation of reef flat of 34.1%. Differently, the seagrasses were clearly identified by the classification and the two sub-classes showed differences in density or species. In this case clear differences in the spatial extent of the seagrass beds were found in comparison with NOAA data back to nearly 15 years.

4. Discussion

4.1. Shoreline, Mangroves, Coral Reefs, and Seagrasses

The difference in shoreline length (Table 2 in results paragraph) is due to the diverse techniques applied for shoreline extraction: manual digitalization (NOAA) and the use of classification algorithms by satellite imagery (ENEA). Furthermore, the wide range of shoreline values could be produced from the different nominal scale of source maps applied in order to extract the coastline [36].

Regarding the mangrove habitat, notwithstanding the different methodologies or imagery applied in NOAA and ENEA studies, the mangrove cover comparison with previous data is interesting. The difference in the extension of the mangroves from 2003–2006 imagery to 2017 is evidenced, but the disparity of NOAA and ENEA habitat mapping methods imposes caution when interpreting this result. The difference equal to 1100 ha between NOAA and ENEA (Table 2 in results paragraph), meaning an increase of the mangrove cover equal to 25.0% in the last decade (2006–2017), is not convincing. The difference between NOAA and ENEA mangrove habitat mapping is likely due to gaps in previous maps. Diachronic maps allow the measurements of temporal changes through concordance and discordance maps. Figure 11 shows a focus of this evidence in the eastern side of the main island of Palau, where it is possible to observe a great concordance between ENEA and NOAA mangrove cover (light green, Figure 11) but also significative discordances like a wide mangrove cover identified only by the present work (dark green, Figure 11) and a very limited mangrove cover identified only by NOAA.

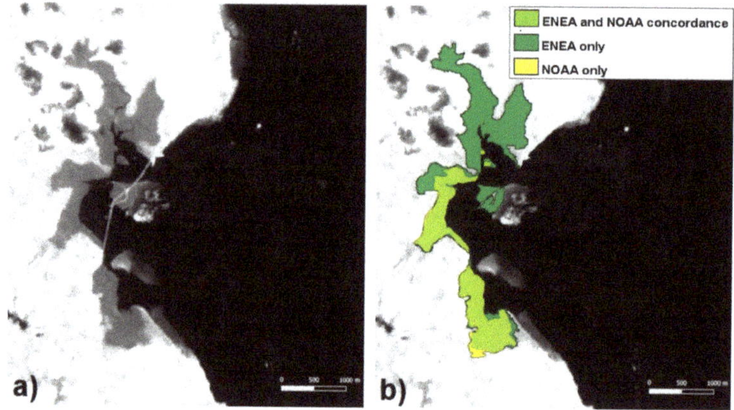

Figure 11. (**a**) A detail of the east sector of the Republic of Palau, with Sentinel-2 enhanced Band 11, where dark grey is the mangrove cover; (**b**) mangrove cover diachronic map showing a concordance wide area of mangroves identified by ENEA and NOAA (light green); two discordance areas with ENEA mangrove cover only (dark green) and NOAA mangrove cover only (yellow), respectively.

On the basis of these results, it is important to note that RS comparisons with different satellite imageries and methods could have some embedded drawbacks. Moreover, the present work's mangrove cover, equal to 5500 ha (Table 2), is in good agreement with 2011 estimations [37]. These latter were achieved with Landsat Enhanced Thematic Mapper (ETM+) data, with images collected during epoch centered on the year 2000, along with very high resolution images such as Ikonos and Quickbird. The overall estimation of 5666 ha for Palau Archipelago [37] is nearly coincident with our results (5500 ha) suggesting an overall equilibrium in Palau Archipelago in the last decades (only 3% difference in hectares in around 17 years). For these reasons an overall equilibrium in the mangrove forests is considered reliable.

With regard to coral reefs and seagrasses, the classification analysis demonstrates that the coral bottom-types are statistically separable and identifiable based on their reflectance spectra. We reason that features in reflectance arise primarily as a result of spectral absorption processes. Radiative transfer modeling shows that in typically clear coral reef waters, dark substrates such as corals have a depth-of-detection limit on the order of 10–20 m [35].

To measure the accuracy of the method proposed on barrier reef and coastal reef zones, we compared results of habitat mapping cover data of main classes from Sentinel-2 image versus data from NOAA maps assumed as reference data. The accuracy varied from 44.3 to 54.6% for the barrier reef estimations and from 39.8 to 56.8% for the coastal reef estimations (Table 4). These results are in the range of overall accuracy estimated for large area studies such as those performed for the Great Barrier Reef (GBR), where overall accuracy estimations of user/producer on Landsat 8 OLI classification for geomorphic zonation and benthic cover types were respectively 59.5 and 32.8% [24]. On the Southern Section Cairn GBR management region, the Sentinel-2 benthic mapping analysis performed on eight reefs with sea truth data validations produced an overall accuracy of 49% using six categories [32].

Reef flat subclasses showed differences between our and NOAA results (Figure 10): two areas were identified by unsupervised classification, but it was impossible to attribute them to turf or algae with different cover. The seagrasses were clearly identified and the cover showed a variation and two further sub-classes probably related to different density or different species were found.

4.2. Limits and Challenges of the Present Approach vs. Similar Studies

In the present work, Palau Island habitat classes/subclasses were chosen with the objective to achieve a RS habitat map centred on the most ecologically important habitats sensitive to climate changes in the Pacific region, like coral reefs, seagrasses, and mangroves. Habitat classes were combined in a unique geomorphological and ecological classification, a method considered as the most appropriate for RS of tropical coastal areas [29]. The unsupervised classification used for benthic habitat classification exploits the advantages of statistical segmentation to find natural boundaries in a dataset and provides consistent classification at multiple sites with little to no ground truth required [30]. However, without a ground validation, it was not possible to map turf, coralline algae, macroalgae, and seagrasses at species level. Differently, the mangrove cover along the island coastline has been identified with high detail, in relation to the high radiometry of the Sentinel-2 spectral signals (14 bit). The method showed its potentiality for medium (seasonal) and long-time (decadal) assessment of changes in sensible communities like seagrasses, also in conjunction with past Landsat 8 imagery collections, but it is limited to main habitat reconnaissance of large-scale habitat mapping [31].

Sensitivity of coral reefs, seagrasses, and mangroves to any effect of climate change varies for the three habitats. It is difficult to separate any effects of climate change from those produced by coastal development and land use practices. In the past six decades, the Palau Archipelago experienced in 1998 and 2010 two episodes of extreme heat associated with El Niño, considered conducive for coral bleaching [38]. Bleaching was most severe in the north-western lagoon, whereas in the bays where temperatures were higher than elsewhere, bleaching and mortality were low. Although the intensity and duration of elevated temperatures are strong predictors of a coral's fate, affecting survival and reproduction [4], the extent of bleaching is different not only by site but it can affect also deep reef. For this reason, the survey of shallow water coral habitats with RS assumes a greater importance as they are source for new propagules also for deeper outer slope coral communities [39]. Seagrasses are likely to be highly sensitive to increases in sea surface temperature, whether they occur as short-term spikes over periods of hours or as chronic exposures for weeks or months. Temperature extremes are known to reduce seagrass growth and lead to plant mortality with reduced carbon sequestration and habitat alteration [4,40]. Increases in seawater temperature also translate into increased disturbance via stronger extreme weather events such as heavy rainfall and tropical cyclones and storms, which put additional stress on seagrass habitats. Direct climate change impacts on mangrove ecosystems are likely to be less significant than the effects of associated sea level rise [2,41]. Sensitivity of mangroves to increased sea surface temperature is likely to be moderate. Rise in temperature and the direct effects of increased CO_2 levels are likely to increase mangrove productivity, change the timing of flowering and fruiting, and expand the ranges of mangrove species into higher latitudes.

5. Conclusions

The proposed approach showed how nowadays the availability of different resources such as free Sentinel-2 imagery and ancillary information allows a reliable classification of marine habitats and the quantification of high value tropical habitats colonized by coral, seagrasses, and mangroves. The accuracy of the method and a revisiting period of the Sentinel-2 satellites of 5–15 days offer the possibility to follow communities covers from season to season and to assess environmental changes over time. Although the value of ground validation is evident to disentangle some uncertainty of interpretation, we demonstrate that the proposed approach is appropriate for extensive large-scale habitat classifications in remote sites like Palau Republic and all Pacific islands. Furthermore, the present methodology can be a good base for future monitoring programmes to be conducted also by resident personnel trained in studies for Marine Spatial Planning and Marine Protected Areas.

Finally, the present work demonstrates the chance offered by free availability of imagery and information to optimize time and resources through worldwide collaboration of research teams to mitigate the effects of climate change in remote Pacific islands.

Supplementary Materials: The following are available online at http://www.mdpi.com/2077-1312/7/9/316/s1, Table S1: Jeffries-Matusita spectral separability coefficients: values are between 1.9–2.0 among the different classes, except the combination "back–reef sand–algae and patch reef coral–sand" with a JM value of 1.8.

Author Contributions: Conceptualization, F.I. and E.C.; Methodology, F.I., A.P., M.B., I.D., E.C., and S.C.; Formal analysis, F.I., E.C., A.P., M.B., I.D., and S.C.; Writing—original draft preparation, A.P., I.D., S.C., and M.B.; Writing—review and editing, A.P., I.D., S.C., and F.I.

Funding: This research received no external funding.

Acknowledgments: We thank N.M. Caminiti for supporting our work, and D. Dominici S. Petric, E. Wang and three anonymous referees whose suggestions greatly improved the manuscript.

Conflicts of Interest: The authors declare no conflict of interest.

References

1. Hoegh-Guldberg, O.; Mumby, P.J.; Hooten, A.J.; Steneck, R.S.; Greenfield, P.; Gomez, E.; Harvell, C.D.; Sale, P.F.; Edwards, A.J.; Caldeira, K.; et al. Coral Reefs Under Rapid Climate Change and Ocean Acidification. *Science* **2007**, *318*, 1737–1742. [CrossRef]

2. Waycott, M.; McKenzie, L.M.; Mellors, J.E.; Ellison, J.C.; Sheaves, M.T.; Collier, C.; Schwarz, A.; Webb, A.; Johnson, J.E.; Payri, C.E. Vulnerability of mangroves, seagrasses and intertidal flats in the tropical Pacific to climate change. In *Vulnerability of Tropical Pacific Fisheries and Aquaculture to Climate Change*; Bell, J.D., Johnson, J.E., Hobday, A.J., Eds.; Secretariat of the Pacific Community: Noumea, New Caledonia, 2011.

3. Hoegh-Guldberg, O.; Poloczanska, E.S.; Skirving, W.; Dove, S. Coral Reef Ecosystems under Climate Change and Ocean Acidification. *Front. Mar. Sci.* **2017**, *4*, 00158. [CrossRef]

4. Smale, D.A.; Wernberg, T.; Oliver, E.C.J.; Thomsen, M.; Harvey, B.P.; Straub, S.C.; Burrows, M.T.; Alexander, L.V.; Benthuysen., J.A.; Donat, M.C.; et al. Marine heatwaves threaten global biodiversity and the provision of ecosystem services. *Nat. Clim. Chang.* **2019**, *306*, 306–312. [CrossRef]

5. Hughes, T.P.; Baird, A.H.; Bellwood, D.R.; Card, M.; Connolly, S.R.; Folke, C.; Grosberg, R.; Hoegh-Guldberg, O.; Jackson, J.B.; Kleypas, J. Climate Change, Human Impacts, and the Resilience of Coral Reefs. *Science* **2003**, *301*, 929–933. [CrossRef]

6. Grottoli, A.G.; Rodrigues, L.J.; Palardy, J.E. Heterotrophic plasticity and resilience in bleached corals. *Nature* **2006**, *440*, 1186–1189. [CrossRef]

7. Wilkinson, C.R.; Buddemeier, R.W. *Global Climate Change and Coral Reefs: Implications for People and Reefs*; Technical Report for the UNEP-IOC-ASPEI-IUCN Global Task Team on the Implications of Climate Change on Coral Reefs; IUCN: Gland, Switzerland, 1994.

8. Mumby, P.J.; Harborne, A.R. Development of a systematic classification scheme of marine habitats to facilitate regional management and mapping of Caribbean coral reefs. *Biol. Conserv.* **1999**, *88*, 155–163. [CrossRef]

9. Jupiter, S.; Roelfsema, C.M.; Phinn, S.R. Science and management. In *Coral Reef Remote Sensing*; Goodman, J.A., Phinn, S.R., Purkis, S., Eds.; Springer: Berlin, Germany, 2013; pp. 403–427.

10. Hedley, J.D.; Roelfsema, C.M.; Chollett, I.; Harborne, A.R.; Heron, S.F.; Weeks, S.; Skirving, W.J.; Strong, A.E.; Eakin, C.M.; Christensen, T.R.L.; et al. Remote Sensing of Coral Reefs for Monitoring and Management: A Review. *Remote Sens.* **2016**, *8*, 118. [CrossRef]

11. Green, E.P.; Mumby, P.J.; Edwards, A.J.; Clark, C.D. A review of remote sensing for the assessment and management of tropical coastal resources. *Coast. Manag.* **1996**, *24*, 1–40. [CrossRef]

12. Phinn, S.R.; Roelfsema, C.M.; Stumpf, R. Remote sensing: Discerning the promise from the reality. In *Integrating and Applying Science: A Handbook for Effective Coastal Ecosystem Assessment*; Longstaff, B.J., Carruthers, T.J.B., Dennison, W.C., Lookingbill, T.R., Hawkey, J.M., Thomas, J.E., Wicks, E.C., Woerner, J., Eds.; IAN Press: Cambridge, MD, USA, 2010.

13. Phinn, S.R.; Hochberg, E.; Roelfsema, C.M. Airborne photography, multispectral and hyperspectral remote sensing on coral reefs. In *Coral Reef Remote Sensing*; Goodman, J.A., Phinn, S.R., Purkis, S., Eds.; Springer: Berlin, Germany, 2013; pp. 3–25.

14. Andréfouët, S.; Claereboudt, M.; Matsakis, P.; Pagès, J.; Dufour, P. Typology of atoll rims in Tuamotu Archipelago (French Polynesia) at landscape scale using SPOT HRV images. *Int. J. Remote Sens.* **2001**, *22*, 987–1004. [CrossRef]

15. Andréfouët, S.; Kramer, P.; Torres-Pulliza, D.; Joyce, K.E.; Hochberg, E.J.; Garza-Pérez, R.; Mumby, P.J.; Riegl, B.; Yamano, Y.; White, W.H.; et al. Multi-site evaluation of IKONOS data for classification of tropical coral reef environments. *Remote Sens. Environ.* **2003**, *88*, 128–143.

16. Naseer, A.; Hatcher, B.G. Inventory of the Maldives coral reefs using morphometrics generated from LANDSAT ETM+ imagery. *Coral Reefs* **2004**, *23*, 161–168. [CrossRef]

17. Yamano, H.; Shimazaki, H.; Matsunaga, T.; Ishoda, A.; McClennen, C.; Yokoki, H.; Fujita, K.; Osawa, Y.; Kayanne, H. Evaluation of various satellite sensors for waterline extraction in a coral reef environment: Majuro Atoll, Marshall Islands. *Geomorphology* **2006**, *82*, 398–411. [CrossRef]

18. Naseer, A.; Hatcher, B.G. Assessing the integrated growth response of coral reefs to monsoon forcing using morphometric analysis of reefs in Maldives. In Proceedings of the 9th International Coral Reef Symposium, Bali, Indonesia, 23–27 October 2000; Volume 1, pp. 75–80.

19. Storlazzi, C.D.; Logan, J.B.; Field, M.E. Quantitative morphology of a fringing reef tract from high resolution laser bathymetry: Southern Molokai, Hawaii. *Geol. Soc. Am.Bull.* **2003**, *115*, 1344–1355. [CrossRef]

20. Giardino, C.; Bresciani, M.; Fava, F.; Matta, E.; Brando, V.E.; Colombo, R. Mapping Submerged Habitats and Mangroves of Lampi Island Marine National Park (Myanmar) from in Situ and Satellite Observations. *Remote Sens.* **2016**, *8*, 2. [CrossRef]

21. Andréfouët, S.; Muller-Karger, F.E.; Robinson, J.A.; Kranenburg, C.J.; Torres-Pulliza, D.; Spraggins, S.A.; Murch, B. Global assessment of modern coral reef extent and diversity for regional science and management applications: A view from space. In Proceedings of the 10th International Coral Reef Symposium, Okinawa, Japan, 28 June–2 July 2004; Volume 2, pp. 1732–1745.

22. Hedley, J.D.; Roelfsema, C.M.; Phinn, S.R.; Mumby, P.J. Environmental and sensor limitations in optical remote sensing of coral reefs: Implications for monitoring and sensor design. *Remote Sens.* **2012**, *4*, 271–302. [CrossRef]

23. Hedley, J.; Roelfsema, C.; Koetz, B.; Phinn, S. Capability of the SENTINEL-2 mission for tropical coral reef mapping and coral bleaching detection. *Remote Sens. Environ.* **2012**, *120*, 145–155. [CrossRef]

24. Hedley, J.D.; Roelfsema, C.; Brando, V.; Giardino, C.; Kutser, T.; Phinn, S.; Mumby, P.J.; Barrilero, O.; Laporte, J.; Koetz, B. Coral reef applications of Sentinel-2: Coverage, characteristics, bathymetry and benthic mapping with comparison to Landsat 8. *Remote Sens. Environ.* **2018**, *216*, 598–614. [CrossRef]

25. Golbuu, Y.; Bauman, A.J.; Kuartei, J.; Victor, S. The State of Coral Reef Ecosystems of Palau. In *The State of Coral Reef Ecosystems of the United States and Pacific Freely Associated States*; Waddell, J., Ed.; NOAA Technical Memorandum NOS NCCOS 11; NOAA/NCCOS Center for Coastal Monitoring and Assessment Biogeography Team: Silver Spring, MD, USA, 2005; pp. 488–507.

26. Kerra, M.J.; Purkisb, S. An algorithm for optically-deriving water depth from multispectral imagery in coral reef landscapes in the absence of ground-truth data. *Remote Sens. Environ.* **2018**, *210*, 307–324. [CrossRef]

27. Hedley, J.D.; Harborne, A.R.; Mumby, P.J. Technical note: Simple and robust removal of sun glint for mapping shallow-water benthos. *Int. J. Remote Sens.* **2005**, *26*, 2107–2112. [CrossRef]

28. Richards, J.A. *Remote Sensing Digital Image Analysis. An Introduction*; Springer: Berlin, Germany, 2006; pp. 1–454.

29. Mumby, P. Methodologies for Defining Habitats. In *Tropical Coastal Management. Habitat Classification and Mapping*; Edwards, A.J., Ed.; UNESCO: Paris, France, 2000; pp. 131–139.

30. Foster, G.; Gleason, A.; Costa, B.; Battista, T.; Taylor, C. Acoustic application. In *Coral Reef Remote Sensing, A Guide for Mapping, Monitoring and Management*; Goodman, J.A., Purkis, S.J., Phinn, S.R., Eds.; Springer: New York, NY, USA, 2013; pp. 221–225.

31. Bouvet, G.; Ferraris, J.; Andrefouet, S. Evaluation of large-scale unsupervised classification of New Caledonia reef ecosystem using Landsat ETM+ imagery. *Oceanol. Acta* **2003**, *26*, 281–290. [CrossRef]

32. NOAA Biogeography Branch. *Shallow-Water Benthic Habitats of American Samoa, Guam, and the Commonwealth of the Northern Mariana Islands: Manual*; NOAA National Ocean Service, National Centers for Coastal Ocean Science: Silver Spring, MD, USA, 2005; p. 33.

33. Analytical Laboratories of Hawaii. *Mapping of Benthic Habitat of Palau*; Technical Report for NOAA; NOAA: Silver Spring, MD, USA, 2007; pp. 1–37.

34. Padma, S.; Sanjeevi, S. Jeffries Matusita based mixed-measure for improved spectral matching in hyperspectral image analysis. *Int. J. Appl. Earth Obs. Geoinf.* **2014**, *32*, 138–151. [CrossRef]

35. Hochberg, E.J.; Atkinson, M.J.; Andrefouet, S. Spectral reflectance of coral reef bottom-types worldwide and implications for coral reef remote sensing. *Remote Sens. Environ.* **2003**, *85*, 159–173. [CrossRef]

36. Weisstein, E.W. Coastline Paradox. From *MathWorld*-A Wolfram Web Resource. Available online: http://mathworld.wolfram.com/CoastlineParadox.html (accessed on 22 August 2019).

37. Bhattarai, B.; Giri, C. Assessment of mangrove forests in the Pacific region using Landsat imagery. *J. Appl. Remote Sens.* **2011**, *5*, 1–11. [CrossRef]

38. Barkley, H.C.; Cohen, A.L.; Mollica, N.R.; Brainard, R.E.; Rivera, H.E.; DeCarlo, T.M.; Lohmann, J.P.; Drenkard, E.J.; Alpert, A.E.; Young, C.W.; et al. Repeat bleaching of a central Pacific coral reef over the past six decades (1960–2016). *Commun. Biol.* **2018**, *1*, 1–10. [CrossRef]

39. Colin, P.L. Ocean warming and the reefs of Palau. *Oceanography* **2018**, *31*, 126–135. [CrossRef]

40. Collier, C.J.; Waycott, M. Temperature extremes reduce seagrass growth and induce mortality. *Mar. Pollut. Bull.* **2014**, *83*, 483–490. [CrossRef]

41. Ellison, J.C. Possible impacts of predicted sea-level rise on South Pacific mangroves. In *Sea-Level Changes and Their Effects*; Noye, J., Grzechnik, M., Eds.; World Scientific: Singapore, 2001; pp. 49–72.

Journal of
*Marine Science
and Engineering*

Article

Statistical Deviations in Shoreline Detection Obtained with Direct and Remote Observations

Giovanni Pugliano [1], Umberto Robustelli [1,*], Diana Di Luccio [2,*], Luigi Mucerino [3], Guido Benassai [1] and Raffaele Montella [2]

[1] Science and Technologies Department, University of Naples "Parthenope", 80143 Napoli, Italy;
    giovanni.pugliano@uniparthenope.it (G.P.); guido.benassai@uniparthenope.it (G.B.)
[2] Engineering Department, University of Naples "Parthenope", 80143 Napoli, Italy;
    raffaele.montella@uniparthenope.it
[3] Department of Earth, Environment and Life Sciences, University of Genova, 16126 Genova, Italy;
    luigi.mucerino@edu.unige.it
* Correspondence: umberto.robustelli@uniparthenope.it (U.R.); diana.diluccio@uniparthenope.it (D.D.L.)

Received: 29 March 2019; Accepted: 6 May 2019; Published: 11 May 2019

Abstract: Remote video imagery is widely used for shoreline detection, which plays a fundamental role in geomorphological studies and in risk assessment, but, up to now, few measurements of accuracy have been undertaken. In this paper, the comparison of video-based and GPS-derived shoreline measurements was performed on a sandy micro-tidal beach located in Italy (central Tyrrhenian Sea). The GPS survey was performed using a single frequency, code, and carrier phase receiver as a rover. Raw measurements have been post-processed by using a carrier-based positioning algorithm. The comparison between video camera and DGPS coastline has been carried out on the whole beach, measuring the error as the deviation from the DGPS line computed along the normal to the DGPS itself. The deviations between the two dataset were examined in order to establish possible spatial dependence on video camera point of view and on beach slope in the intertidal zone. The results revealed that, generally, the error increased with the distance from the acquisition system and with the wash up length (inversely proportional to the beach slope).

Keywords: DGPS measurements; video camera observation; shoreline position; beach survey

---

1. Introduction

Coastal areas are highly dynamic environments that provide important benefits, but are also subject to a variety of natural hazards such as beach erosion, tsunamis [1,2], and floods [3]. For coastal zone monitoring and coastal risk assessment [4], shoreline detection is a fundamental work. The shoreline is the line where land meets the sea, and due to the dynamic nature of the sea, sometimes it is difficult to determine a precise line that can be called the "shoreline". As reported by Boak and Turner [5], a functional definition of the "shoreline", which has to consider the shoreline in both a temporal and spatial sense, is required.

Many authors highlighted the existence of different methodologies for coastal monitoring [6–9], not only limited to shoreline detection, based on direct and remote acquisition systems.

Direct shoreline surveys are normally carried out using the GPS technique by post-processing or real-time methods [10]. The main limitation of this method is the huge time required for covering large stretches of the coastline and the difficulties inherent in doing ad hoc timely post-storm measurements.

Remote shoreline observations can be distinguished in remote sensing [11–13], UAV (Unmanned Aerial Vehicles) [14], and video monitoring [15,16], which were first introduced by Aarninkhof [17] and Turner et al. [18]. These remote observations have been also extensively used to validate wind and wave numerical models [19–23] and to outline the environmental big picture in marine spatial planning applications [24–26].

In addition to using UAVs in beach survey [27,28], video monitoring can provide a remotely-sensed measurement, fixed at a secure location (e.g., a tower or high-point), with the capability of acquiring imagery at a frequency ranging from fractions of seconds to hours. The technology is relatively low-cost, but the main issue is the processing method, especially the image rectification process, considering that the imagery is strongly oblique and relies on a number of GCPs (Ground Control Points) for finding the best geometry solutions. This technique has been successfully applied for both shoreline monitoring [29,30], rip current measurement [14] and morphodynamics classification of sandy beach [31,32].

Several techniques are proposed to extract shoreline from images, based on discriminating sea from sand. Plant and Holman [33] used a method initially developed for gray-scale cameras, called Shoreline Intensity Maximum (SLIM), which defines the shoreline as the cross-shore position at which wave breaking is maximized and consequently corresponds to a maximum in pixel intensity. With the adoption of color cameras, spectral information could also be used to identify the shoreline, making use of the water property to absorb the Red signal (R) and the sand property to absorb the Green (G) and blue signal [34,35]. Following Almar et al. [36], we identify beach pixels as channel values with a high R/G ratio, whereas water pixels as high G channel values. The shoreline represents the transition zone between the two peaks. In order to smooth out high-frequency signals that are caused by disturbances like foam, we used time-exposure images [17,18].

In this paper, we performed shoreline detection by means of remote video camera observations and DGPS direct shoreline measurements on a sandy microtidal beach located in central Tyrrhenian Sea. Almar et al. [36], among other authors, have presented video-based methods to determine shoreline position in different tidal and wave energy contexts, showing an error dependence on local swash length, which is inversely proportional to the beach slope. Following this research item, we first performed a preliminary analysis of the GPS survey, which was followed by the evaluation of the statistical deviations of the video camera coastline measurements from the DGPS ones. The errors have been spatially processed, in order to examine if systematic deviations can be ascribed only to the intertidal beach slope or to additional factors, like the distance from the video camera. In particular, we examined the longitudinal and cross-shore errors associated with the video camera point of view.

The paper is structured as follows: after a brief description of the study area (Section 2), we illustrate the survey and validation methods used in Section 3. The results with the relative discussion are reported in Sections 5 and 6, respectively.

2. Study Area

The study area includes the beach of Serapo, located in the Central Thyrrenian Sea (Italy), which exhibits a rather regular morphology and rectilinear shoreline that is complexly long, approximately 1.45 km (Figure 1), aligned in the NW-SE direction (for the statistical processing, we considered only a 1.34 km-long shoreline, directly sampled with GPS). This beach is characterized by homogeneous grain size features with median diameter around 0.38 mm, indicating medium sands [37]. The topographic survey data, acquired in September 2017, refer to coastline and to five beach profiles T1–T5 as reported in Table 1. Coastline investigations were conducted in September 2018 (for the location, see Figure 1). Based on the difference in beach width and berm height, the studied beach can be subdivided into the following two stretches: a western beach stretch including profiles T1 and T2, which is 45–60 m wide and characterized by lower berm height (1.10–1.30 m) and a mean emerged beach slope of 5.5%; an eastern beach stretch including profiles T3, T4, and T5, which is characterized by higher beach width of 74–87 m and higher berm height of 1.29–1.46 m, with a lower emerged beach slope of 1.48–3.94%. This different beach width was already explained by Di Luccio et al. [37] in terms of partial clockwise beach rotation around a central pivotal point.

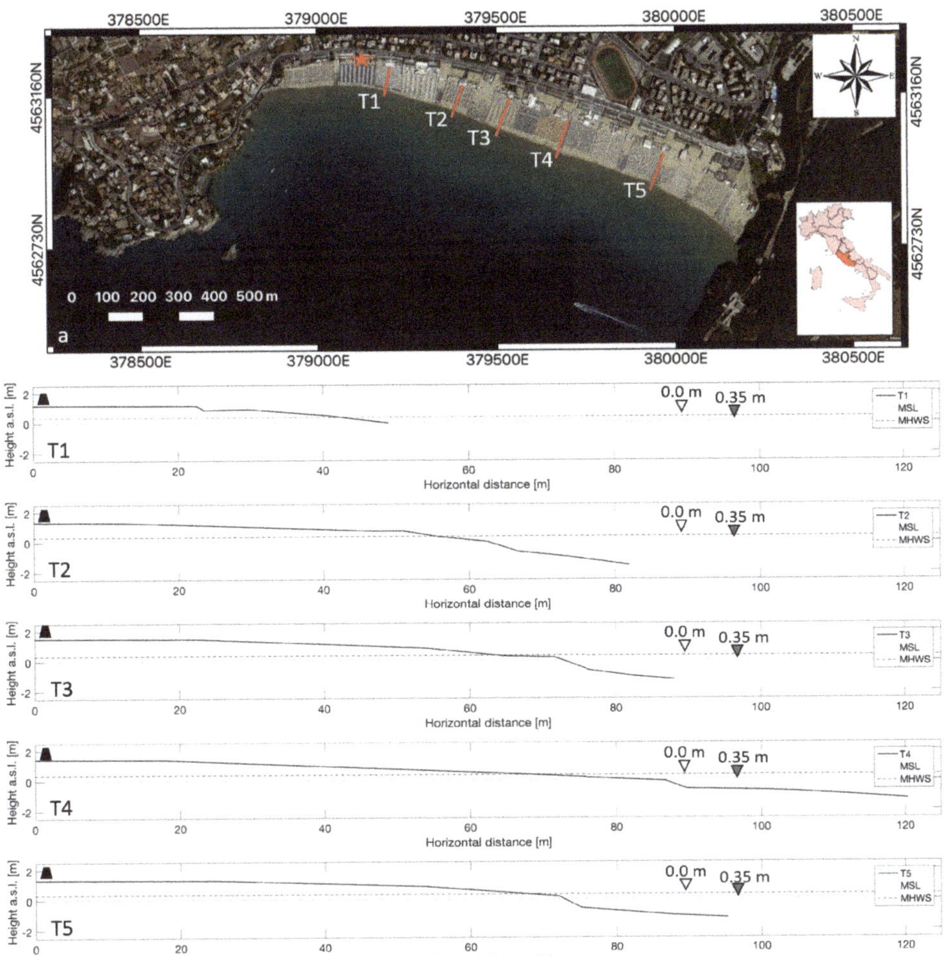

Figure 1. Serapo beach study area (**a**) with the location of the beach video camera system (red star) and the five investigated cross-shore transects (red lines), obtained by a beach survey conducted in September 2017; the beach profile along the transect with the location of the anthropic structures (black trapezium), Mean Sea Level (MSL), and the Mean Spring Tidal Range (MHSW), extracted by the official Italian tide archives (http://www.mareografico.it, last access: 30 April 2019) was reported for T1, T2, T3, T4, and T5. (basemap ©2018 Google product).

Table 1. Summary of Serapo beach's morphological characterization. The reported parameters are relative to the beach cross-shore profiles T1–T5.

Profile	Emerged Beach width L (m)	Berm Height (m)	Emerged Beach Slope β (%)	Local Beach Slope γ (%)
T1	45.38	1.11	7.88	5.85
T2	61.21	1.29	3.19	15.02
T3	73.54	1.46	3.94	17.27
T4	86.79	1.37	1.48	17.87
T5	74.02	1.29	3.88	21.37

3. Materials and Methods

3.1. Sea Waves' Analysis

The wave climate is connected to the methodology used for the definition of the coastline since the morphological changes affecting the beach after a sea storm can change the longitudinal beach profile (beach slope) in a more or less accentuated manner, influencing, as we shall see then, the remote acquisition processes. Moreover, if storm patterns and wave distributions (time scales) change, the coastline shape can evolve with erosion/accretion processes [38], inducing possible beach rotation, well evident with continuous video monitoring systems.

In order to define the sea condition characterizing the study area, we analyzed the historical time series extracted by the Ponza buoy 3-hourly data, provided by the Italian Sea Wave Measurement Network [39]. This directional pitch-roll buoy is moored in deep water [40], a few miles to the south of Ponza island ($40°52'00.10''$ N, $12°56'60.00''$ E). The available parameters are significant wave height H_s, mean wave period T_m, and mean wave direction D_m, covering the years 1989–2014 (with some data gaps). Climatological wave analysis is presented, highlighting the wave storms that exposed a lower limit threshold of H_s equal to 2 m.

3.2. Kinematic GPS Survey

The Global Positioning System (GPS) surveying was employed to get the reference shoreline position. The collection of shore-parallel GPS positions was carried out in September 2017 using a single-frequency code and carrier phase receiver (Trimble Pathfinder ProXH) as the rover. The most common shoreline detection technique applied to visibly-discernible shoreline features is manual visual interpretation in the field, as reported by Boak and Turner [5]. In this paper, the shoreline was compiled by interpolating between described shore-normal beach profiles and a series of points collected along the beach face, which included maximum run-up limit and 0.5-m depth.

In order to compute the positioning solutions by various modes including single-point and relative positioning, post-processing of the raw data was performed using the RTKLIB software [41], an open-source software for standard and precise positioning [41]. The processing option for the relative positioning mode was set to carrier-based kinematic positioning; the ambiguity resolution was performed both recalculating the phase bias estimates every epoch (instantaneous) and using the Kalman filter to estimate the phase biases over many epochs (continuous). The last solution was much less noisy, as expected, and was adopted as the reference line. The analysis of the single point and carrier-based instantaneous solutions was included as well. In particular, the use of the single-point positioning was investigated as having the potential for being a low-cost method for shoreline mapping. In order to improve the results obtainable with this technique, the researchers have directed their studies towards the identification and reduction of multipath [42–44] as it is the highest source of error in the single-point positioning. In the last few months, smart phones equipped with a dual frequency multi-constellation GNSS receiver have appeared on the market. They could represent a valid and low-cost alternative to carry out the survey as shown by Robustelli et al. [45]. The position accuracy was evaluated on the basis of the calculation of the distance and azimuth of the estimated horizontal position error vector (Figure 2).

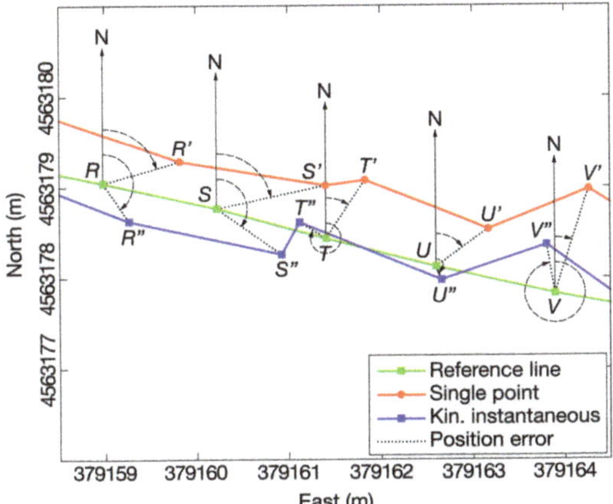

Figure 2. Scheme of the quantification of errors in the kinematic GPS horizontal positions referred to a portion of the trajectory.

Further analysis was carried out to discern also the effect of random GPS errors on the trajectory of the kinematic path. The magnitude of the effect due to random errors was identified through the computation of the deflection angle given by the angle between a V′W′ line and the prolongation of the preceding U′V′ line (Figure 3). Deflection angles were computed as positive or negative values depending on whether the line lied to the right (clockwise) or left (counterclockwise) of the prolongation of the preceding line.

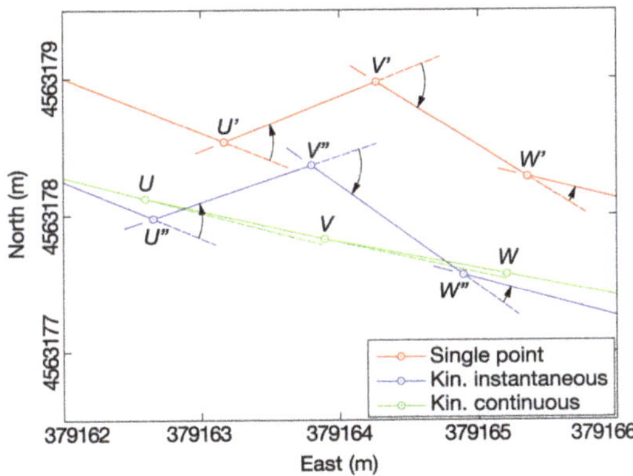

Figure 3. Scheme of the deflection angles referred to a portion of the trajectory.

3.3. Video Camera Observations

A video monitoring system was installed in the western part of Serapo beach (41°12′41.81″ N–13°33′29.66″ E) as shown in Figure 4. Two cameras provided a total view of the bay with 1920 ×1080 pixel resolution from an elevation of about 11 m above Mean Sea Level (MSL) and 100 m from the coastline. The mean pixel resolution was evaluated using the methodology suggested in Holland et al. [46]. Images were collected every second from 08:00 until 16:00 local time on a local video-recorder with a 2-TB storage capability from May 2017–June 2018. Optical measurements are subject to image distortions due to inherent camera characteristics [47]. According to Stumpf et al. [48] and Mucerino et al. [29], the images were calibrated using Ground Control Points (GCPs), which were placed in the view area of the cameras and were acquired in UTM-WGS84 using DGPS. The calibration process required at least nine GCPs for each camera that covered the entire camera view; camera system calibration was performed by using 32 GCPs: 21 on the southeast side camera and 11 on west side camera. Finally, the image dataset was processed using Beachkeeper plus software [15].

Shoreline detection from the image was based on the physical consideration that the color contrast between beach and water is sufficient, lighting is strong enough, and the number of pixels in the water and beach groups is sufficient. Following Almar et al. [36] and other less recent authors [49–52], in order to reduce errors due to sea level variations, the shoreline was detected using the swash signature on timex average images, using images averaged over short periods (30 s), which significantly improved the accuracy of shoreline determination. The shoreline thus identified on the oblique image was converted to real-world coordinates using ortho-rectification techniques. Finally, shorelines obtained from corresponding images from the two cameras were managed to form a whole continuous shoreline.

Figure 4. Camera system located on Serapo beach: (**a**) camera records, southeast side of the bay; (**b**) camera detections, west side of Serapo shoreline; (**c**) a detail of the camera installation.

3.4. Validation Method

The comparison between video and DGPS coastline has been performed on the whole beach, measuring the error as the deviation from the continuous DGPS line computed, depicted in green, along the normal to the DGPS itself for a sample of about 500 points. These points were chosen keeping constant the mutual distance (about 2.7 m) along the track.

In this paper, an error model for video camera measurements is proposed. The displacement offset with respect to the normal direction to the coastline can be computed by an exact trigonometric relation. In Figure 5, let ε_N represent the error in the normal direction to the coastline, ε_L the longitudinal error along the line of sight from the camera to the object, and ε_T the transverse error, perpendicular to the line of sight. The angle α is the angle between the coastline and the line of sight. If α has been observed, then:

$$\varepsilon_N = \varepsilon_L \sin \alpha + \varepsilon_T \cos \alpha \tag{1}$$

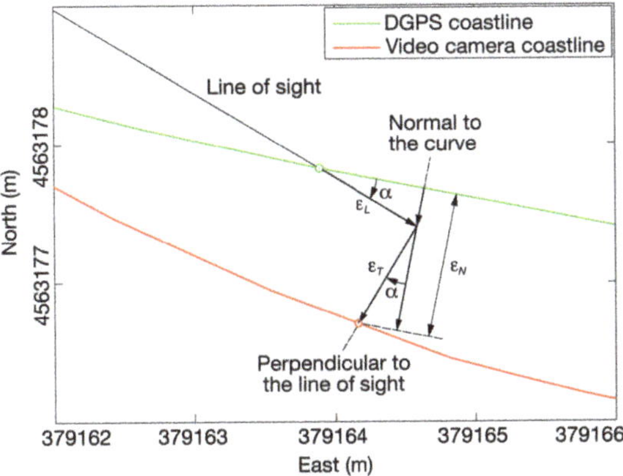

Figure 5. Scheme of the longitudinal and transverse error projections along the normal direction to the coastline.

The error model was based on two error components, the longitudinal ε_L and the transverse ε_T error, which were projected in the direction of the coastline's normal. The longitudinal and transverse errors were computed from the following equation:

$$\varepsilon_L = \varepsilon_T = aD - be^{cs} \tag{2}$$

where D is the distance of the shoreline from the video camera, s is the intertidal beach slope, e is Napier's constant, and a, b, c are three constants to be determined experimentally. In Equation (2), the correction be^{cs} may be applied to the distance-dependent error aD as the intertidal beach slope increases, which is in agreement with the inverse proportionality of the normal error with the beach slope [36].

Substituting Equation (2) into Equation 1 and if the longitudinal and transverse errors are grouped, this yields:

$$\varepsilon_N = (aD - be^{cs})(\sin\alpha + \cos\alpha) \tag{3}$$

4. Results

4.1. Wave Climate Conditions

Significant wave height H_s, mean wave period T_m, and mean wave direction D_m, covering the years 1989–2014 (with some data gaps) are reported in Figure 6a–c, respectively. Maximum values of H_s for the ten highest wave storms recorded in the observation period are reported in Table 2, together with the mean period T_m, peak period T_p, and mean direction D_m associated with the storm peak. The highest H_s value was observed in December 1999, in agreement with Piscopia [53]. The selection of the ten highest storms showed that the directions of the storm waves were confined between 238° N and 272° N. This result is consistent with the one obtained with the selection of all the storms ($H_s > 2$ m) highlighted in red in Figure 6, showing that the highest waves were associated with southwestern, western, and northwestern directions [37]. With regard to astronomical sea level variations, the study area experiences a typical semi-diurnal tide, with a mean tidal range of 0.35 m, following the official Italian tide archives (http://www.mareografico.it, last access: 30 April 2019). However, main sea level variations due to meteorological surges can reach values up to 1 m [54].

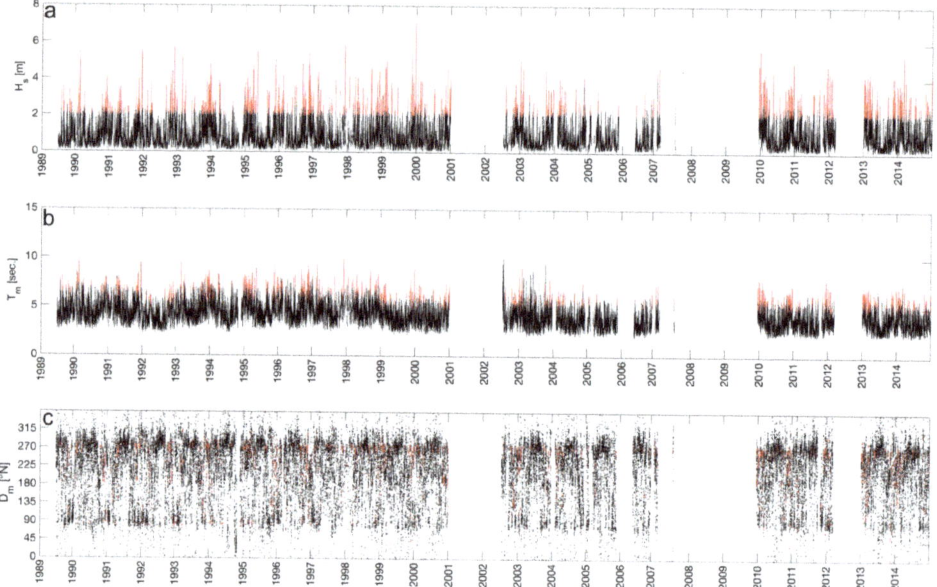

Figure 6. Offshore wave parameters (H_s in (**a**), T_m in (**b**), and D_m in (**c**)) recorded by the Ponza buoy during the years 1989–2014 with three-hourly time steps. The red color highlights the wave parameters associated with the highest waves ($H_s > 2$ m).

Table 2. Summary of the ten highest wave storm events recorded by the Ponza buoy in the period 1989–2014.

Storm Id	H_s max (m)	T_m (s)	T_p (s)	D_m (°N)	Date and Hour (UTC)
1	7.10	8.7	12.5	266.0	28 December 1999 15:00
2	5.80	10.0	11.1	272.0	3 December 1997 15:00
3	5.70	10.5	12.5	271.0	28 February. 1990 03:00
4	5.70	7.5	10.0	262.0	6 December 1992 03:00
5	5.61	7.9	10.5	264.6	2 January 2010 06:00
6	5.60	9.4	11.1	272.0	20 December 1991 15:00
7	5.50	8.9	10.0	266.0	13 May 1995 21:00
8	5.50	8.7	10.0	275.0	21 November 1996 06:00
9	5.50	7.7	10.0	238.0	19 November 1999 09:00
10	5.35	7.4	8.7	256.8	4 March 2014 06:00

4.2. Comparison between Different Accuracies of GPS Solutions

The results of the comparison of the single point and kinematic instantaneous GPS solutions with the continuous one are shown below.

As shown in Figure 7, the errors of the horizontal positions obtained in single-point mode reached a maximum value of 2.45 m, with a mean horizontal error of 0.93 m (Figure 7a). It should be noted that this average value was greater than the average error of the kinematic instantaneous solutions (Figure 7b) equal to 0.49 m. The standard deviation of the single point position errors was, however, comparable to that of the kinematic case, with a value of 0.28 m. Figure 7a also denotes a clear systematic error in terms of direction with prevailing azimuths towards northeast.

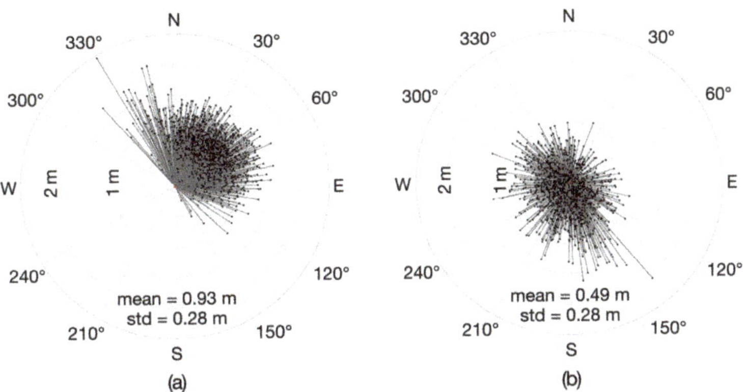

Figure 7. Single-point (**a**) and kinematic instantaneous (**b**) horizontal position error.

From the polar histogram of the azimuths shown in Figure 8, it is noted that about 85% of the single point error directions fell in the first quadrant from 0–90 degrees, with about 65% of the directions concentrated between the azimuth of 15 degrees and the azimuth of 75 degrees. This systematic error was reduced for the directions of the kinematic instantaneous errors; in fact, 85% of the errors were distributed in a range of greater amplitude, which extended from 0–240 degrees.

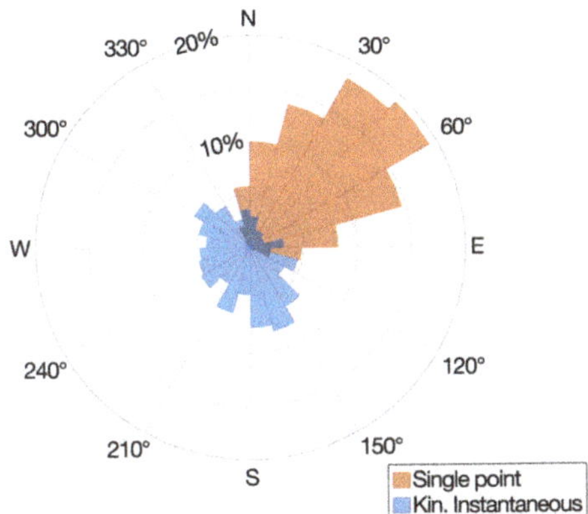

Figure 8. Histogram of the single-point and kinematic instantaneous error position azimuths.

With regard to the deflection angle analysis reported in the Methods, we computed its standard deviation, which was much lower for the kinematic continuous trajectory with respect to the single point and kinematic instantaneous ones, as reported in Figure 9. This circumstance confirms the capability of the kinematic continuous option for smoothing GPS data due to the use of Kalman filter.

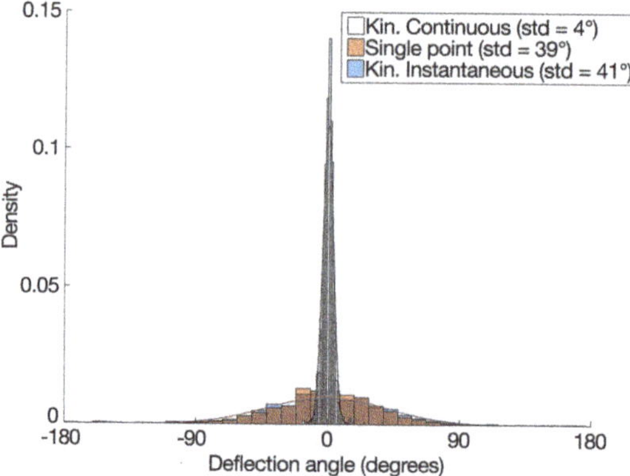

Figure 9. Histogram of the deflection angles.

4.3. Comparison between DGPS and Video Camera Coastline

In Figure 10, the comparison between the video camera coastline (in green) and the DGPS one (in red) is reported for four adjacent west-east beach sectors. The error distribution is not only dependent on the camera distance, as explained in the following.

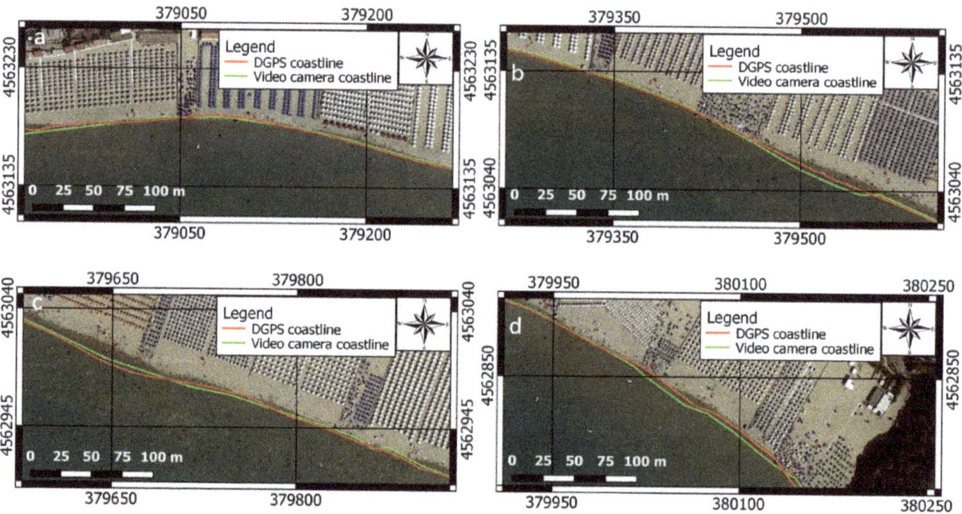

Figure 10. Video camera coastline (in green) compared with the DGPS coastline (in red) along Serapo beach. (basemap ©2018 Google product). In order to guarantee a better representation of the two coastlines, the beach has been divided into four sections, shown in the panels **a**–**d**, starting from west (panel **a**) to east (panel **d**).

In Figure 11a, we report in the chromatic scale the deviation between the two coastlines along the beach. Moreover, in Figure 11b, we report the error magnitude, neglecting the sign, as a function of the shoreline length. The following results can be synthesized.

First, the error exhibited a different amount between points located at a distance lower or higher than approximately one third of the beach extent, as shown from Figure 11b. In particular, for the eastern camera images, from the shoreline length of 200 m to a distance of 500 m, the error was around 1 m, while for the more distant points, the error was higher, with a maximum value of 5 m.

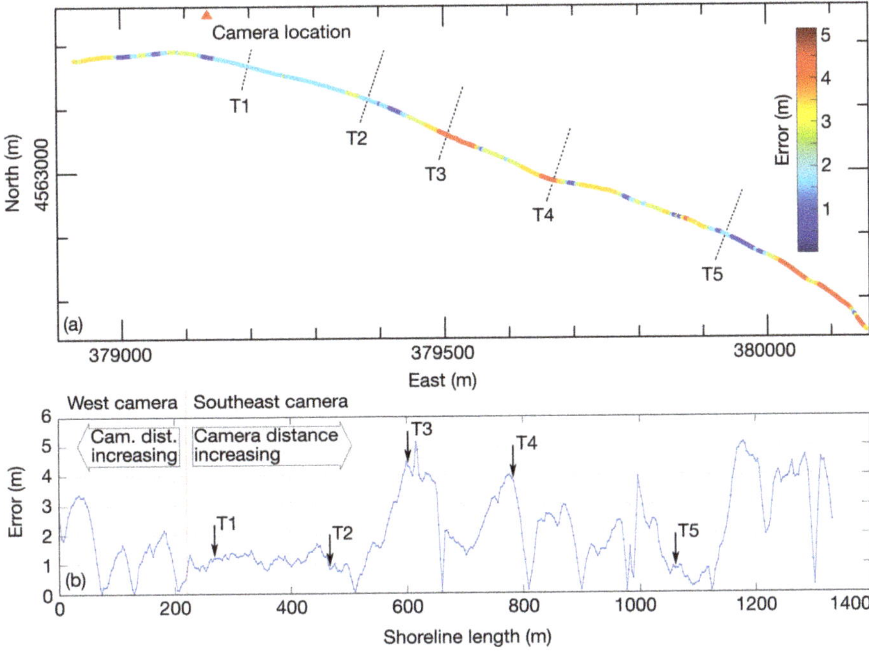

Figure 11. Observed deviations of the video coastline in the normal direction to the DGPS line for about 500 sample points. Panel a shows the deviations in the chromatic scale, in panel b the deviations are reported as function of the shoreline length.

Secondly, the error showed a lower standard deviation for the closer points; this is evident from the quite uniform color in Figure 11a if compared with the color variation along the second part of the coastline. The summary of the error statistics is reported in Table 3.

Table 3. Summary of the error statistics.

Distance (m)	Mean (m)	Max (m)	std (m)
0–200	1.62	3.37	0.97
200–500	1.08	1.69	0.29
500–1340	2.40	5.13	1.34

The inspection of the deviations between the remotely-measured coastline and the direct DGPS ones showed a lower error amount for points located at a distance closer than one third of the beach extent. This result has been obtained both in terms of maximum error (less than 2.0 m) and for the error standard deviation (less than 0.3 m). On the contrary, the more distant points exhibited a maximum error up to 5.0 m and an error standard deviation up to 1.3 m.

5. Discussion

In this paper, a shoreline detection technique from a low-cost video monitoring station was used, applying an automatic extraction technique that undertakes a water/beach distinction from color band ratios and a shoreline slope recognition module.

Prior to the validation of the video-derived shoreline with the direct survey, the differences between the single point and the unfiltered kinematic GPS have been examined. The results showed that the unfiltered kinematic GPS positions exhibited a better performance of the single-point GPS with respect to the systematic error. Nevertheless, the latter presented a comparable standard deviation and a reduced cost both in terms of instruments and operational speed.

Compared to a direct DGPS survey, the video-derived coastline acquisition was less time consuming and more cost effective. The possibility to acquire a beach topography with a high temporal frequency can potentially highlight coastal processes during the winter season, when a direct survey is difficult due to harsh weather conditions.

To perform the error analysis, as stated previously (Equation (3)), it is useful to break the normal error into a distance-dependent part ε'_N and a slope-dependent one ε''_N.

The distance-dependent part can be estimated as:

$$\varepsilon'_N = aD \left(\sin \alpha + \cos \alpha \right) \tag{4}$$

where the constant a is computed experimentally from the field data; the values were 0.0125 and 0.0079 for the west side camera (from 0–200 m) and the southeast side camera (from 200–1340 m), respectively. Thus, the distance-dependent error is represented in Figure 12, with positive (negative) values indicating seaward (landward) offsets of the shoreline. The light blue curve represents normal errors that increase with the distance compared to the observed values (red curve). Figure 12 shows results obtained when the model is not corrected for the beach slope.

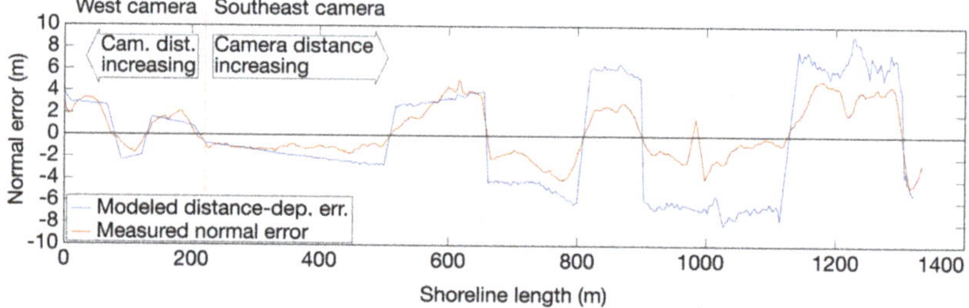

Figure 12. Distance-dependent part of the normal error.

In performing normal error estimation, it should be assumed that there are possible corrections due to the increasing intertidal beach slope. The model adopted to estimate these corrections is:

$$\varepsilon''_N = be^{cs} \left(\sin \alpha + \cos \alpha \right) \tag{5}$$

where the constants b and c are computed experimentally from the field data and the slope s is expressed in percent units. The constants b, c and the slope correction ε''_N have been calculated only for the stretch of coast between Transect 1 and Transect 5 for which the slope values have been obtained by interpolation. The values were 0.005 and 1/3 for the constants b and c, respectively. Figure 13 shows the relationship between the slope and the correction due to the beach slope.

Substituting the appropriate values into Equation (3), yields:

$$\varepsilon_N = \left(0.0079D - 0.005e^{\frac{s}{3}} \right) \left(\sin \alpha + \cos \alpha \right) \tag{6}$$

The result depends on the combination of three components including camera distance D, beach slope s, and the angle α between the coastline and the line of sight. The error model for video camera measurements has been tested to the extent between Transect 1 and Transect 5 due to the field data not enabling calculation of the beach slope and determining the error offset for the entire study area. Figure 14 shows the estimated normal errors determined by the final computed Equation (6). The particular combination produced results corresponding to those observed with the exception of the value at the shoreline length of 600 m.

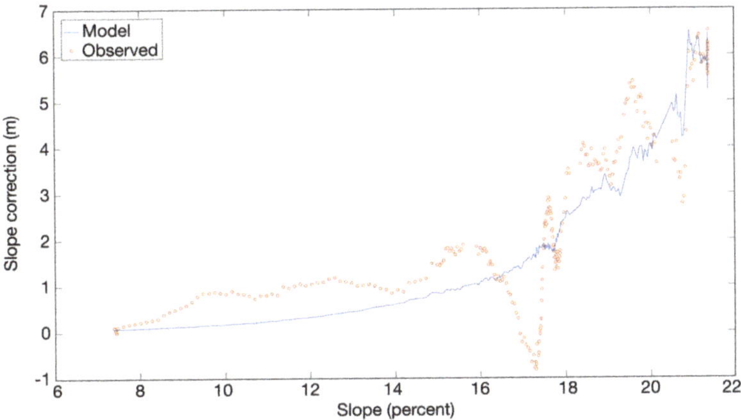

Figure 13. Slope correction due to the increasing intertidal beach slope between Transects 1 and 5.

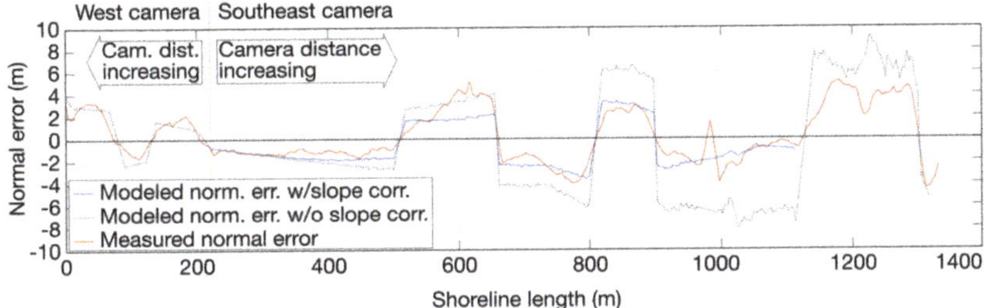

Figure 14. Estimated normal errors.

6. Conclusions

Currently, risk assessment plays a fundamental role in preventing irreversible erosion processes, as well as flooding damages. In this context, video monitoring represents a low-cost instrument to store a large amount of data with a high temporal frequency irrespective of weather conditions.

In this paper, we analyzed the limitations of the video-based coastline in terms of distance from the point of view and influence of beach topography. A shoreline detection technique from a low-cost video monitoring station was validated against DGPS-derived shoreline.

We introduced an error model, based on the transverse and longitudinal error along the line of sight, and computed a correction factor that can be applied to the distance-dependent error.

Wave storms alter the amounts of wave energy approaching a shore and can change the beach cross-shore profile, influencing the applicability of video monitoring acquisition systems.

Several types of coastal information can be provided from images. In Di Luccio et al. [16], we analyzed wave run-up; in the near future, we will use data for rip-current detection, which we already analyzed with direct and remote observation tools [14].

Day-light limitations of the video-derived data can be overcome by thermal cameras, which can operate also during night hours in order to monitor the beach during severe events. This goal will be the continuation of the present research, together with the challenge of modeling new algorithms to lower the deviation from direct measurements.

Author Contributions: Conceptualization G.P., G.B., U.R., D.D.L., L.M. Supervision G.B. Methodology G.P., G.B., U.R., D.D.L., L.M. Formal Analysis G.P., U.R., D.D.L. Validation G.P., G.B., U.R., D.D.L. Visualization G.P. and D.D.L. Software R.M. Data Curation R.M. Investigation, L.M. Writing—Original Draft Preparation G.P., U.R. and D.D.L. Writing—Review and Editing G.P., U.R. and D.D.L.

Funding: The research activity has been supported by the Parthenope University of Naples with a grant within the call "Support for Individual Research for the 2015–17 Period". The above support is gratefully acknowledged.

Acknowledgments: The authors are grateful to the forecast service of the University of Napoli *"Parthenope"* (http://meteo.uniparthenope.it) for the HPC facilities and to the beach club "Nave di Serapo" (www.navediserapo.it) for having hosted the video monitoring equipment.

Conflicts of Interest: The authors declare no conflict of interest.

References

1. De Girolamo, P.; Di Risio, M.; Romano, A.; Molfetta, M. Landslide tsunami: Physical modeling for the implementation of tsunami early warning systems in the Mediterranean Sea. *Procedia Eng.* **2014**, *70*, 429–438. [CrossRef]

2. Samaras, A.; Karambas, T.V.; Archetti, R. Simulation of tsunami generation, propagation and coastal inundation in the Eastern Mediterranean. *Ocean Sci.* **2015**, *11*, 643–655. [CrossRef]

3. Small, C.; Nicholls, R.J. A global analysis of human settlement in coastal zones. *J. Coast. Res.* **2003**, *19*, 584–599.

4. Di Risio, M.; Bruschi, A.; Lisi, I.; Pesarino, V.; Pasquali, D. Comparative analysis of coastal flooding vulnerability and hazard assessment at national scale. *J. Mar. Sci. Eng.* **2017**, *5*, 51. [CrossRef]

5. Boak, E.H.; Turner, I.L. Shoreline definition and detection: A review. *J. Coast. Res.* **2005**, *21*, 688–703. [CrossRef]

6. Bruno, M.F.; Molfetta, M.G.; Pratola, L.; Mossa, M.; Nutricato, R.; Morea, A.; Nitti, D.O.; Chiaradia, M.T. A Combined Approach of Field Data and Earth Observation for Coastal Risk Assessment. *Sensors* **2019**, *19*, 1399. [CrossRef] [PubMed]

7. Bruno, M.F.; Molfetta, M.G.; Mossa, M.; Morea, A.; Chiaradia, M.T.; Nutricato, R.; Nitti, D.O.; Guerriero, L.; Coletta, A. Integration of multitemporal SAR/InSAR techniques and NWM for coastal structures monitoring: Outline of the software system and of an operational service with COSMO-SkyMed data. In Proceedings of the 2016 IEEE Workshop on Environmental, Energy, and Structural Monitoring Systems (EESMS), Bari, Italy, 13–14 June 2016; pp. 1–6.

8. Holman, R.A.; Stanley, J. The history and technical capabilities of Argus. *Coast. Eng.* **2007**, *54*, 477–491. [CrossRef]

9. Valentini, N.; Saponieri, A.; Molfetta, M.G.; Damiani, L. New algorithms for shoreline monitoring from coastal video systems. *Earth Sci. Inform.* **2017**, *10*, 495–506. [CrossRef]

10. Benassai, G.; Di Luccio, D.; Mucerino, L.; Paola, G.D.; Rosskopf, C.M.; Pugliano, G.; Robustelli, U.; Montella, R. Shoreline rotation analysis of embayed beaches in the Central Thyrrenian Sea. In Proceedings of the 2018 IEEE International Workshop on Metrology for the Sea; Learning to Measure Sea Health Parameters (MetroSea), Bari, Italy, 8–10 October 2018; pp. 7–12. [CrossRef]

11. Guariglia, A.; Buonamassa, A.; Losurdo, A.; Saladino, R.; Trivigno, M.L.; Zaccagnino, A.; Colangelo, A. A multisource approach for coastline mapping and identification of shoreline changes. *Ann. Geophys.* **2006**, *49*, 295–304.

12. Alesheikh, A.A.; Ghorbanali, A.; Nouri, N. Coastline change detection using remote sensing. *Int. J. Environ. Sci. Technol.* **2007**, *4*, 61–66. [CrossRef]

13. Nunziata, F.; Buono, A.; Migliaccio, M.; Benassai, G. Dual-polarimetric C-and X-band SAR data for coastline extraction. *IEEE J. Sel. Top. Appl. Earth Obs. Remote Sens.* **2016**, *9*, 4921–4928. [CrossRef]

14. Benassai, G.; Aucelli, P.; Budillon, G.; De Stefano, M.; Di Luccio, D.; Di Paola, G.; Montella, R.; Mucerino, L.; Sica, M.; Pennetta, M. Rip current evidence by hydrodynamic simulations, bathymetric surveys and UAV observation. *Nat. Hazards Earth Syst. Sci.* **2017**, *17*, 1493–1503. [CrossRef]

15. Brignone, M.; Schiaffino, C.F.; Isla, F.I.; Ferrari, M. A system for beach video-monitoring: Beachkeeper plus. *Comput. Geosci.* **2012**, *49*, 53–61. [CrossRef]

16. Di Luccio, D.; Benassai, G.; Budillon, G.; Mucerino, L.; Montella, R.; Pugliese Carratelli, E. Wave run-up prediction and observation in a micro-tidal beach. *Nat. Hazards Earth Syst. Sci.* **2018**, *18*, 2841–2857. [CrossRef]

17. Aarninkhof, S.G.J. Nearshore Bathymetry Derived From Video Imagery. Ph.D. Thesis, Delft University of Technology, Delft, The Netherlands, 2003.

18. Turner, I.L.; Aarninkhof, S.; Dronkers, T.; McGrath, J. CZM applications of Argus coastal imaging at the Gold Coast, Australia. *J. Coast. Res.* **2004**, *20*, 739–752. [CrossRef]

19. Montuori, A.; Ricchi, A.; Benassai, G.; Migliaccio, M. Sea wave numerical simulation and verification in Tyrrhenian costal area with X-band cosmo-skymed SAR data. In Proceedings of the ESA, SOLAS & EGU Joint Conference Earth Observation for Ocean-Atmosphere Interactions Science, Frascati, Italy, 29 November–2 December 2011; Volume 29.

20. Benassai, G.; Montuori, A.; Migliaccio, M.; Nunziata, F. Sea wave modeling with X-band COSMO-SkyMed© SAR-derived wind field forcing and applications in coastal vulnerability assessment. *Ocean Sci.* **2013**, *9*, 325–341. [CrossRef]

21. Benassai, G.; Migliaccio, M.; Montuori, A.; Ricchi, A. Wave simulations through SAR COSMO-SkyMed wind retrieval and verification with buoy data. In Proceedings of the Twenty-second International Offshore and Polar Engineering Conference. International Society of Offshore and Polar Engineers, Rhodes, Greece, 17–22 June 2012.

22. Dominici, D.; Zollini, S.; Alicandro, M.; Della Torre, F.; Buscema, P.M.; Baiocchi, V. High Resolution Satellite Images for Instantaneous Shoreline Extraction Using New Enhancement Algorithms. *Geosciences* **2019**, *9*, 123. [CrossRef]

23. Palazzo, F.; Latini, D.; Baiocchi, V.; Del Frate, F.; Giannone, F.; Dominici, D.; Remondiere, S. An application of COSMO-Sky Med to coastal erosion studies. *Eur. J. Remote Sens.* **2012**, *45*, 361–370. [CrossRef]

24. Benassai, G.; Di Luccio, D.; Corcione, V.; Nunziata, F.; Migliaccio, M. Marine Spatial Planning Using High-Resolution Synthetic Aperture Radar Measurements. *IEEE J. Ocean. Eng.* **2018**, *43*, 586–594. [CrossRef]

25. Di Tullio, G.R.; Mariani, P.; Benassai, G.; Di Luccio, D.; Grieco, L. Sustainable use of marine resources through offshore wind and mussel farm co-location. *Ecol. Model.* **2018**, *367*, 34–41. [CrossRef]

26. Benassai, G.; Di Luccio, D.; Migliaccio, M.; Cordone, V.; Budillon, G.; Montella, R. High resolution remote sensing data for environmental modeling: Some case studies. In Proceedings of the 2017 IEEE 3rd International Forum on Research and Technologies for Society and Industry (RTSI), Modena, Italy, 11–13 September 2017; pp. 1–5.

27. Nunziata, F.; Buono, A.; Migliaccio, M.; Benassai, G.; Di Luccio, D. Shoreline erosion of microtidal beaches examined with UAV and remote sensing techniques. In Proceedings of the 2018 IEEE International Workshop on Metrology for the Sea; Learning to Measure Sea Health Parameters (MetroSea), Bari, Italy, 8–10 October 2018; pp. 162–166. [CrossRef]

28. Di Luccio, D.; Benassai, G.; Di Paola, G.; Rosskopf, C.; Mucerino, L.; Montella, R.; Contestabile, P. Monitoring and modeling coastal vulnerability and mitigation proposal for an archaeological site (Kaulonia, Southern Italy). *Sustainability* **2018**, *10*, 2017. [CrossRef]

29. Mucerino, L.; Albarella, M.; Carpi, L.; Besio, G.; Benedetti, A.; Corradi, N.; Firpo, M.; Ferrari, M. Coastal exposure assessment on Bonassola bay. *Ocean Coast. Manag.* **2019**, *167*, 20–31. [CrossRef]

30. Schiaffino, C.F.; Dessy, C.; Corradi, N.; Fierro, G.; Ferrari, M. Morphodynamics of a gravel beach protected by a detached low-crested breakwater. The case of Levanto (eastern Ligurian Sea, Italy). *Ital. J. Eng. Geol. Environ.* **2015**, *15*, 31–39.

31. Lisi, I.; Molfetta, M.; Bruno, M.; Di Risio, M.; Damiani, L. Morphodynamic classification of sandy beaches in enclosed basins: The case study of Alimini (Italy). *J. Coast. Res.* **2011**, 180–184.

32. Postacchini, M.S.L.C.; Mancinelli, A. Medium-term dynamics of a middle Adriatic barred beach. *Ocean Sci.* **2017**, *3*, 719. [CrossRef]

33. Plant, N.G.; Holman, R.A. Intertidal beach profile estimation using video images. *Mar. Geol.* **1997**, *140*, 1–24. [CrossRef]

34. Bryan, K.R.; Smith, R.; Ovenden, R. The Use of a Video Camera to Assess Beach Volume Change During 2001 at Tairua, New Zealand. In *Coasts & Ports 2003 Australasian Conference, Proceedings of the 16th Australasian Coastal and Ocean Engineering Conference, the 9th Australasian Port and Harbour Conference and the Annual New Zealand Coastal Society Conference*; Institution of Engineers: Canberra, Australia, 2003; p. 1236.

35. Smith, R.; Bryan, K. Monitoring beach face volume with a combination of intermittent profiling and video imagery. *J. Coast. Res.* **2007**, *23*, 892–898. [CrossRef]

36. Almar, R.; Ranasinghe, R.; Sénéchal, N.; Bonneton, P.; Roelvink, D.; Bryan, K.R.; Marieu, V.; Parisot, J.P. Video-based detection of shorelines at complex meso–macro tidal beaches. *J. Coast. Res.* **2012**, *28*, 1040–1048.

37. Di Luccio, D.; Benassai, G.; Di Paola, G.; Mucerino, L.; Buono, A.; Rosskopf, C.M.; Nunziata, F.; Migliaccio, M.; Urciuoli, A.; Montella, R. Shoreline Rotation Analysis of Embayed Beaches by Means of In Situ and Remote Surveys. *Sustainability* **2019**, *11*, 725. [CrossRef]

38. Slott, J.M.; Murray, A.B.; Ashton, A.D.; Crowley, T.J. Coastline responses to changing storm patterns. *Geophys. Res. Lett.* **2006**, *33*. [CrossRef]

39. Bencivenga, M.; Nardone, G.; Ruggiero, F.; Calore, D. The Italian data buoy network (RON). *Adv. Fluid Mech. IX* **2012**, *74*, 321.

40. Arena, F.; Pavone, D. Return period of nonlinear high wave crests. *J. Geophys. Res. Ocean.* **2006**, *111*. [CrossRef]

41. Takasu, T. RTKLIB Ver. 2.4.2 Manual. Available online: http://www.rtklib.com/prog/manual_2.4.2.pdf (accessed on 23 November 2018).

42. Pugliano, G.; Robustelli, U.; Rossi, F.; Santamaria, R. A new method for specular and diffuse pseudorange multipath error extraction using wavelet analysis. *GPS Solut.* **2016**, *20*, 499–508. [CrossRef]

43. Robustelli, U.; Pugliano, G. GNSS code multipath short time fourier transform analysis. *Navi* **2018**, *65*, 353–362. [CrossRef]

44. Robustelli, U.; Pugliano, G. Code multipath analysis of Galileo FOC satellites by time-frequency representation. *Appl. Geomat.* **2018**, *11*, 69–80. [CrossRef]

45. Robustelli, U.; Baiocchi, V.; Pugliano, G. Assessment of Dual Frequency GNSS Observations from a Xiaomi Mi 8 Android Smartphone and Positioning Performance Analysis. *Electronics* **2019**, *8*, 91. [CrossRef]

46. Holland, K.T.; Holman, R.A.; Lippmann, T.C.; Stanley, J.; Plant, N. Practical use of video imagery in nearshore oceanographic field studies. *IEEE J. Ocean. Eng.* **1997**, *22*, 81–92. [CrossRef]

47. Didier, D.; Bernatchez, P.; Augereau, E.; Caulet, C.; Dumont, D.; Bismuth, E.; Cormier, L.; Floc'h, F.; Delacourt, C. LiDAR validation of a video-derived beachface topography on a tidal flat. *Remote Sens.* **2017**, *9*, 826. [CrossRef]

48. Stumpf, A.; Augereau, E.; Delacourt, C.; Bonnier, J. Photogrammetric discharge monitoring of small tropical mountain rivers: A case study at Rivière des Pluies, Réunion Island. *Water Resour. Res.* **2016**, *52*, 4550–4570. [CrossRef]

49. Aarninkhof, S. Argus-based monitoring of intertidal beach morphodynamics. In Proceedings of the Coastal Sediments 99, Long Island, NY, USA, 21–23 June 1999.

50. Alexander, P.S.; Holman, R.A. Quantification of nearshore morphology based on video imaging. *Mar. Geol.* **2004**, *208*, 101–111. [CrossRef]

51. Davidson, M.; Aarninkhof, S.; Van Koningsveld, M.; Holman, R. Developing coastal video monitoring systems in support of coastal zone management. *J. Coast. Res.* **2006**, *I*, 49–56.

52. Holman, R.; Stanley, J.; Ozkan-Haller, T. Applying video sensor networks to nearshore environment monitoring. *IEEE Pervasive Comput.* **2003**, *2*, 14–21. [CrossRef]

53. Piscopia, R.; Inghilesi, R.; Panizzo, A.; Corsini, S.; Franco, L. Analysis of 12-year wave measurements by the Italian Wave Network. In *Coastal Engineering 2002: Solving Coastal Conundrums*; World Scientific: Singapore, 2003; pp. 121–133.

54. Cazenave, A.; Bonnefond, P.; Mercier, F.; Dominh, K.; Toumazou, V. Sea level variations in the Mediterranean Sea and Black Sea from satellite altimetry and tide gauges. *Glob. Planet. Chang.* **2002**, *34*, 59–86. [CrossRef]

MDPI

St. Alban-Anlage 66

4052 Basel

Switzerland

Tel. +41 61 683 77 34

Fax +41 61 302 89 18

www.mdpi.com

Journal of Marine Science and Engineering Editorial Office

E-mail: jmse@mdpi.com

www.mdpi.com/journal/jmse